Konrad Hahm
Die Kunst in Finnland

SEVERUS

Hahm, Konrad: Die Kunst in Finnland
Hamburg, SEVERUS Verlag 2014

ISBN: 978-3-86347-964-0
Druck: SEVERUS Verlag, Hamburg, 2014
Nachdruck der Originalausgabe von 1933

Der SEVERUS Verlag ist ein Imprint der Diplomica Verlag GmbH.

Bibliografische Information der Deutschen Nationalbibliothek:
Die Deutsche Nationalbibliothek verzeichnet diese Publikation in der
Deutschen Nationalbibliografie; detaillierte bibliografische Daten sind im
Internet über http://dnb.d-nb.de abrufbar.

© SEVERUS Verlag
http://www.severus-verlag.de, Hamburg 2014
Printed in Germany
Alle Rechte vorbehalten.

Der SEVERUS Verlag übernimmt keine juristische Verantwortung oder
irgendeine Haftung für evtl. fehlerhafte Angaben und deren Folgen.

KONRAD HAHM

Die Kunst in Finnland

SEVERUS

VORWORT

Finnland hat in der Reihe der nordischen Länder immer etwas abseits der größeren europäischen Zusammenhänge gestanden. Seine Zugehörigkeit zum russischen Reiche hat es mit sich gebracht, daß Finnland für Europa kein selbständiger Begriff war. Erst durch den heldenmütigen Kampf, den das finnische Volk um die Mitte des vorigen Jahrhunderts gegen die russische Unterdrückung zu führen begann, wurde die Aufmerksamkeit der Welt auf diese Nation gerichtet, die in unbeugsamer Freiheitsliebe ihre Selbständigkeit verteidigte. Und als im Weltkrieg in einem Befreiungskampf, an dem auch deutsche Truppen mitwirkten, Finnland sich von Rußland löste, erwies es sich, daß hier ein kulturell und politisch durchgebildetes Volk und auch ein längst vorhandenes Staatsgebilde als viertes skandinavisches Reich in die Geschichte eintrat. Finnland ist in den 15 Jahren seiner Selbständigkeit ein wichtiger Faktor der Politik, Wirtschaft und Bildung Europas geworden. Bei dem immer stärkeren Interesse, das sich in Deutschland diesem neuen Staat und seiner nationalen Kultur zuwendet, wird gerade die Frage nach der bildenden Kunst in Finnland oft gestellt, von der im Gegensatz zu der Kunst Schwedens, Norwegens und Dänemarks allzuwenig bekannt ist. Bisher ist der Versuch einer zusammenfassenden Darstellung der finnischen Kunst in Deutschland noch nicht gemacht worden.

Die vorliegende Schrift will sich dieser Aufgabe annehmen. Sie will einen kurzen Überblick über das Wesen der bildenden Kunst in Finnland geben und will auch dem Finnlandreisenden eine erste Einführung in sie vermitteln.

Bei den vorbereitenden Arbeiten wurde der Verfasser in liebenswürdigster Weise von vielen Gelehrten und Kunstfreunden in Finnland unterstützt. Er dankt besonders dem Auswärtigen Amt in Helsinki, Herrn Minister Wäinö Wuolijoki, Gesandter Finnlands in Berlin, Herrn Prof. Dr. Manninen, Direktor des Finnischen Nationalmuseums, Helsinki, Herrn Dr. Stjernschantz, Direktor des Ateneums, Helsinki, Herrn Dr. Kurt Antell, Herrn Architekt Prof. Jussi Paatela, Herrn Stadtarchitekt Gunnar Taucher, Herrn Dr. Wennervirta in Helsinki.

INHALT

EINLEITUNG 5
    Land und Volk 5
    Volk und Kunst 7

DAS MITTELALTER 8

NACH DER REFORMATION 11

SEIT 1800 15
    Die neuere Baukunst 17
    Die Malerei 22
    Die Plastik 30

VOLKSKUNST UND KUNSTHANDWERK 33

Die finnischen Ortsnamen werden (wie alle finnischen Worte) auf der ersten Silbe betont. Hinter den finnischen Ortsnamen ist, soweit gebräuchlich, die schwedische Benennung angegeben.

EINLEITUNG

LAND UND VOLK

Finnland, in der Landessprache Suomi genannt, die Landbrücke von Skandinavien nach Europa und Asien, zwischen dem 60. und 70. nördlichen Breitengrad, im Süden von der Ostsee, im Norden vom Eismeer umspült, durchschnitten vom Polarkreis, enthält in Land und Volk die Mischung von Härte und Weichheit, von Kälte und Wärme, die den Norden kennzeichnet. Hunderttausende von Schären, von Inseln und Klippen umsäumen die Meeresküsten, riesige Seengebiete in unendlicher Weite und Mannigfaltigkeit mit unzähligen Inseln durchziehen das Land von Süden nach Norden, wo die Stromschnellen und Wasserfälle, die großen Wälder, Moore und Einöden beginnen. Eine gewaltige Symphonie von Gegensätzen, von Stein und Wald, von Land und Wasser, von Licht und Dunkelheit gestaltet das Bild des Landes. Die Mitternachtssonne und die Nordlichter des Winters, der erst im Mai die Eisdecke der Seen freigibt, die silbernen Lichtnächte des Sommers, in denen die Vegetation schnell wächst wie im Rausch, die schweigsame Tiefe von Wäldern und Seen, die helle Birke und die dunkle Kiefer, die Kargheit des Bodens, der doch eine tropische Fülle von Waldbeeren hervorbringt, der Auerhahn und das Renntier, der Bär und die Schwalbe, das Moos und der Obstbaum sind die Sinnbilder seiner Natur, die Nördliches mit Südlichem, Östliches mit Westlichem vereint. Sie hat die gesteigerte Schärfe Norwegens ohne seine Gebirgsmonumentalität und Rauheit, die Lieblichkeit südschwedischer Provinzen gepaart mit tieferem Ernst, die Warmheit mecklenburgischer Landschaften, gemischt mit fast südlicher Romantik von Klippen und Buchten und mit der musikalischen Schwermut östlicher Weite und Verlorenheit.

Das Volk, das hier lebt, ist eigenartig und kräftig wie das Land, willensstark, zäh, lebhaft und schweigsam. Von einem alten Volkstum getragen, das von der finnisch-ugrischen Völkergruppe abgezweigt und in engster Auseinandersetzung und Mischung mit germanischen und baltischen Stämmen zu einer besonderen Volkseinheit geworden ist, hat sich dieses Bauernvolk in seiner Isoliertheit stark und jung erhalten, nicht nur in seiner Sprache, sondern auch in seinem Wesen. Der finnische Bauer, überwiegend ein Kleinbauer, (auch heute ist nur wenig Großgrundbesitz vorhanden), ist nie Leibeigener gewesen. Seine Wesensart ist aus dieser Bauernfreiheit herausgewachsen und kennzeichnet den finnischen Volkscharakter: Freiheitsliebe und Ordnungssinn, Beharrlichkeit und Vorsicht und ein starkes Rechts- und Selbstbewußtsein, das mit einer ebenso starken Wahrheitsliebe und einer sprichwörtlichen Ehrlichkeit verbunden ist. Erzogen von der Natur und dem abgeschlossenen Volkstum, ist der Finnländer nicht nur ein fest umrissener Volkstyp, sondern auch infolge seiner guten Anlagen und seiner Begabung ein Kulturmensch. Wir haben ja gerade durch die nordischen Völker gelernt, daß das Kulturniveau nicht abhängig ist von Marmor, Gold und Seide, von Weltbeherrschung und Luxus. Der finnische Dichter Juhani Aho kennzeichnet seine Volksgenossen folgendermaßen:

„Es ist durchaus keine Laune des Zufalls, daß gerade wir Finnen hier in Finnland geblieben sind und uns da bis heute gehalten haben. Hierher sind auch andere gekommen, die nach Land suchten. Aber sie sind entweder durch den Hof durchgefahren oder haben an der Pforte kehrt gemacht. ... Auch die Finnen hätten mildere Länder gefunden, wo Milch und Honig fließt. Aber ihre Neigung scheint sie nach immer rauheren Ländern geführt zu haben. Grade wie zum Trotz haben sie dürre Heiden, Sümpfe und dichte Wälder aufgesucht, wo nie der Frost aus der Erde weicht.

Manche meinen, daß sie Stärkeren auszuweichen gezwungen gewesen wären. Ich glaube, daß das ihr praktischster Königsgedanke gewesen ist. Sie haben die Zähigkeit ihrer Lenden gekannt, und dort, wo anderer Leute Rückgrat gebrochen wäre, dort hat ihr Rückgrat sich nur gekräftigt. Die Hacke ist ihr Schwert gewesen, und mit ihr haben sie für sich das Land bezwungen, das auch der Eroberer als ihr eigen anerkannt hat. Wie er auch geheißen haben mag, sprach er Schwedisch, Dänisch oder Russisch, das Endergebnis aus den Kämpfen zwischen den Eroberern ist das gewesen, daß der Kampfplatz den Finnen selbst geblieben ist. Den ausländischen Ansiedlern ist es eine allzu harte Nuß gewesen. Und so ist es auch noch. Wenn wir es dem Fremden anböten und sagten: „Hier — komm und nimm", so wäre niemand da, es zu nehmen.

Deswegen können wir hier ziemlich sorgenfrei sitzen und ohne uns zu beunruhigen, ‚was der Tag morgen bringt'. Wir können ruhig auf das Rauschen der himmlischen Winde lauschen, wie der Wacholder auf steinigem Hügel. Der einschlagende Blitz zermalmt die Urwaldstanne, aber in den Wacholderbusch sinkt er kraftlos. Kriegsrosse stampfen über ihn hin, die Lafetten der Kanonen pressen ihn an den Boden. Aber der Wacholder bricht nicht. Er bekommt keine blutende Wunde, keinen Knochenbruch. Ist der Aufruhr vorüber, streckt der kleine Baum seinen kurzen, sehnigen Körper, und Ast spricht zu Ast: ‚Du wachse dorthin, ich wachse hier'. Und es dauert nicht lange, bis die Spur der Füße und die Furche der Räder verwachsen ist. Und wenn der, der darüber fuhr, morgen seine gestrigen

Spuren sucht, findet er sie nicht mehr. Der Weg ist zugewachsen, und der Wacholderbusch scheint unberührt.

Grade, daß wir die härteste Felszunge gewählt haben, wo nur, uns gleichgeartet, der Wacholder wachsen kann, das war unsere größte Weisheit. Der Moses, der uns hierher leitete, verstand die eigentümliche Kraft, die sich in seines Volkes Wacholdernatur birgt."

In dem dünn besiedelten Lande, das Fruchtboden zwischen Wald (der 60% der Bodenfläche bedeckt), See und Feld nur streckenweise freigibt, ist die übliche Siedlungsform das in weiten Abständen voneinander gelegene Einzelgehöft (Abb. 28-31), seltener (meist als Kirchdorf) die Dorfanlage. Diese Isoliertheit hat das Gemeinschaftsgefühl der Bevölkerung stark ausgebildet, auf dem Gebiete des modernen Genossenschaftswesens ist Finnland heute führend, es hat darin bei dem Mangel großer Vermögen eine starke volkswirtschaftliche Grundlage gefunden. Bei aller Beharrlichkeit ist dieses noch überwiegend bäuerliche Land (70% Landwirtschaft) doch höchst fortschrittlich eingestellt. Neben dem Ruderboot steht das Auto; neben der Sauna, der alten finnischen Badestube, die ein Nationalbesitz ursprünglichster Art geblieben ist, ein vorzügliches Verkehrswesen, die kleinste Bahnhofswirtschaft zeugt von der allgemeinen nordischen Eß- und Lebenskultur; den Puukko, das finnische Dolchmesser, trägt der Arbeiter wie der Minister am Gürtel; die allgemeine Volksbildung ist überraschend, auf dem tiefsten Lande hat auch der kleine Bauer neben seinem Motorboot eine Bücherei. Die Teilnahme der Frau am öffentlichen Leben ist schon immer sehr stark gewesen, Finnland führte als erstes Land der Welt 1906 das Frauenstimmrecht ein. Die Liebe zum Sport, eine wirkliche Volksangelegenheit in Finnland, hat eine Reihe von internationalen Namen hervorgebracht.

Auch die Kunst ist ein charakteristisches Stück seiner nationalen Kultur.

VOLK UND KUNST

Aus den Elementen ursprünglichen Volkstums formt sich in Finnland auch die bildende Kunst. Der Norden hat ja stets den Kunstwellen, die von der Mittelmeerkultur und den europäischen Kulturzentren her mit den Stilepochen dort eindrangen, ein eigenartiges und eigenwilliges Gepräge verliehen. Die nordischen Länder, die in viel größeren Zeitabschnitten von dem Wandel der Formen und Ideen erreicht wurden, haben viel weniger davon angenommen als Mittel -und Südeuropa. Ihre Entlegenheit, ihre wirtschaftliche Armut haben stets den konservativen Charakter und das Maß der eigenen Aufnahmefähigkeit hervortreten lassen. Das Eigene und das Fremde stehen hier einander schroffer gegenüber. Wo das Fremde und Neue Eingang findet, wird es dem bestehenden Eigenen angepaßt. Dieser Vorgang der Anpassung ist ja charakteristisch für die Wandlung, die das Formgut der hohen Kunst auch in der Volkskunst erfährt. Das Ergebnis ist immer eine Verringerung der neuen führenden Erscheinungsformen, eine Abkürzung und Umwertung, die vom ästhetischen und stilgeschichtlichen Standpunkt oft als eine Entwertung erscheint. Deshalb hat die Kunstgeschichte diese Minderung oft als Mangel an Gestaltungsvermögen und als Barbarismus angesehen. Gerade im Norden mit seiner hochstehenden Volkskultur würde diese Folgerung oft zu irrigen Anschauungen führen, wenn man nicht berücksichtigt, daß die seelische und nicht die materielle Handhabung das Entscheidende ist. Die nordischen Länder mit ihrer starken Volkskunst haben die Einflüsse der übernationalen europäischen Kunst immer mit den Grundmotiven ihrer bäuerlichen Formgestaltung durchsetzt. Eine Darstellung der nordischen Kunst wird daher in viel stärkerem Maße als eine Geschichte der deutschen, französischen oder italienischen Kunst jene Elemente beachten müssen, die aus der Volkskunst her, ja aus der frühgeschichtlichen Gebundenheit, in Erscheinung und Wirkung getreten sind.

DAS MITTELALTER

Die erste Berührung Finnlands mit der europäischen Kunst vermittelte die Kirche. Finnland wurde in drei Kreuzzügen 1157, 1249 und 1293 von Schweden her erobert und christianisiert. Der englische Mönch Henrik, 1148 zum Bischof von Upsala ernannt, begann unter König Erik dem Heiligen das Bekehrungswerk und erlitt von der Hand des Bauern Lalli den Märtyrertod. Sein prächtiger 1370 in der St. Henriks-Kirche von Nousiainen (Nousis) an seinem Todesort aufgestellter Sarkophag aus schwarzem Tonschiefer mit gravierten Bronzeplatten, wahrscheinlich eine flandrische Arbeit, die das Leben des Märtyrers und den Kampf der Schweden mit den Finnen schildern (Abb. 1), gehört zu den schönsten Sehenswürdigkeiten mittelalterlicher Kunst in Finnland. Das Christentum brachte seine große kulturelle Waffe, den Kirchenbau, ins Land. Die ältesten Kirchen Finnlands werden vermutlich der vorhandenen Holzbautradition entsprochen haben. Es sind jedoch keine Beispiele erhalten, die, wie etwa die norwegischen Stabkirchen, über die finnische Holzarchitektur Aufschluß geben könnten. Die ältesten erhaltenen Kirchenbauten sind Steinkirchen aus der Mitte des 13. Jhdts. (Abb. 4). Ihre Form ist ohne reiche Gliederung die eines hohen Hauses mit einem gewaltigen Giebel und steilem Satteldach. Dieser Typ, der auch im schwedischen Upland und Roslagen anzutreffen ist und sich von den Ålandsinseln her über West- und Mittelfinnland verbreitet, bildet gewöhnlich im Grundriß ein regelmäßiges Rechteck mit einer dreischiffigen Gliederung, ohne einen ausgebildeten Chor an der Ostseite, der nur manchmal durch reichere Gestaltung des Gewölbes und besonderen malerischen Schmuck hervorgehoben wird. Charakteristisch sind zwei Anbauten: die an der Nordseite in der Nähe des Chores befindliche Sakristei und an der Südseite das sogenannte Waffenhaus, die Eingangshalle, in welcher die Männer vor dem Gottesdienst ihre Waffen ablegten.

Die Nordseite weist eine geschlossene Wand auf, um den bösen Mächten der im Norden gelegenen Hölle keinen Einlaß zu bieten, die Südseite ist die Fensterseite. Der Turm ist nur bei den älteren Bauten in Åland und an der Westküste unmittelbarer Teil der Kirche, meist steht er als Glockenturm, oft auch als Torbau für den Friedhofseingang, daneben. Das Baumaterial ist überwiegend Moränenfels, aus dem die weißverfugten Steinmauern in Hohltechnik (je eine Innen- und Außenseite, die Hohlräume gefüllt mit Mörtel, Steinen und Kies) errichtet wurden. Der Ziegel wird nur im Innern bei den Pfeilern und Gewölbekonstruktionen verwendet oder bei der Ornamentierung der Giebel. Diese monumentalen Giebel sind meist mit einem riesigen Kreuz und nischenartigen Ornamenten oder Kreisen (in der Mauerung ausgespart) geschmückt.

Im Gegensatz zu der herben und strengen Sparsamkeit an Schmuck und Gliederung, die sich im Äußeren dieser mächtigen Kirchenhäuser ausprägt, steht das Innere, das mit einer ungewöhnlich reichen, sprühenden Bemalung geschmückt ist (Abb. 5-7). Diese Malereien, in der Reformationszeit der strafenden Übertünchung verfallen, sind erst vor wenigen Jahrzehnten entdeckt und vom finnischen Altertumsverein systematisch freigelegt worden. Es sind Kalkmalereien in der al-secco Technik, mit Wasserfarben auf einen trockenen Kalkgrund aufgetragen. Im 15. Jahrhundert beginnt diese fantastische Belebung der Innenräume, die noch von der nordischen Tierornamentik gespeist zu sein scheint, besonders unterstützt durch die kunstliebenden Bischöfe Magnus Olofson Tavast (1408-50) oder Konrad Bitz (1460-89) und wohl durch den Dominikanerorden, der in Turku (Åbo) und Wiipuri (Wiborg) Klöster hatte und sich im Norden auch der Kunstpflege angenommen hat. In verschwenderischer Fülle breiten sich Bänder, Ranken und Blumen über die Säulen und Gewölbe aus, die Wände und breiten Gewölbeflächen sind mit Schilderungen aus dem alten und neuen Testament, aus dem Leben Christi, der Apostel und Heiligen nach den beliebten Vorlagen der Biblia Pauperum bedeckt, oft bis 400 Darstellungen in einer Kirche, wobei der Chor meist die Bilder der Kirchenväter und der Evangelisten enthält. Die (größtenteils anonymen) Schöpfer dieser Kirchenmalereien sind nicht nur Finnländer, wie wohl Petrus Henriksson, der Kirchenmaler von Uusikirkko (Nykyrko) und seine Schüler, gewesen, oder Schweden, sondern wahrscheinlich auch viel Deutsche, worauf starke Ähnlichkeiten mit der norddeutschen Kirchenmalerei in Rostock, Doberan, Danzig usw. hinweisen.

Diese gotische Kirchenmalerei ist das große eigenartige und glanzvolle Vermächtnis der künstlerischen Kultur Finnlands im Mittelalter. Die interessantesten Beispiele davon sind in den Kirchen zu Rauma (Raumo), Lohja (Lojo), Kumlinge, Taivassalo (Töfsala), Sauvo (Sagu), Hattula, Hollola, Uusikirkko (Nykyrka), Sipoo (Sibbo), Inkoo (Ingå), Kaarina (St. Karins), Finnström, Porvoo (Borgå), Perniö (Bjerno), Helsinge, Nousiainen (Nousis), Laitila (Letala).

Von mittelalterlichen Skulpturen ist nur wenig erhalten, das in ähnlicher Weise wie die Wandmalereien große und eigenartige Gestaltungskraft aufweist, so die Muttergottes aus Pernaja (Pernå), die aus der Mitte des 14. Jhdts. stammt und Zeugnis einer starken heimischen Schnitzkunst ist (Abb. 11 a), oder ein Altarschrein aus Urjala (Urdiala) (Abb. 8), der hundert Jahre älter ist, eins der ältesten Stücke dieser seltenen Gattung der Flügelaltäre mit quadratischem Grundriß überhaupt. Die kirchliche Schnitzerei des 14. und 15. Jahrhunderts zeigt den Einfluß der Hansezeit, die Verbindung mit norddeutschen Werkstätten, mit Lübeck, Hamburg, Danzig, mit Köln und den Niederlanden, wie der etwa von 1425-1450 von dem Hamburger Meister Francke gemalte Flügelaltar mit der Barbaralegende (Abb. 12 u. 13) aus Uusikirkko.

Stärker, weil sie sich auf die vorhandene Volkskunst stützen konnte, weist die kunsthandwerkliche Schnitzerei des Mittelalters eigenes Gepräge auf an einer Reihe von Chorstühlen, Emporen, Geländern und Bänken mit deutlicher Anlehnung an die nordische Kerbschnittradition wie in den Kirchen von Sauvo, Korpo, Uusikirkko, Taivassalo und Hollola (Abb. 14 u. 15). Auch von schmiedeeisernen Arbeiten sind bemerkenswerte Zeugnisse einheimischer Überlieferung vorhanden, wie die Grabgitter in der Domkirche von Turku (Åbo).

Der Dom zu Turku (Abb. 2) weicht vom Typ der finnischen Landkirchen ab. Er gehört in die Reihe der norddeutschen Dome des Mittelalters, als ein Denkmal für die geistige Einheit des Bauwillens der Hansezeit. Seine Vorbilder stehen in Reval, Riga, Danzig (Marienkirche) und Lübeck. Im Jahre 1300 der Jungfrau Maria und dem heiligen Henrik geweiht, hat er durch Brände und Zerstörung vielfach seine Gestalt verändert, die er durch einen großen Umbau 1370 erhielt, und deren eindruckvollster Teil das die Seitenschiffe stark überhöhende Mittelschiff ist (Abb. 3). Der jetzige Hochchor wurde in protestantischer Zeit unter Gustav Adolfals Stiftung eines Turkuer Kaufmanns errichtet. Er ist zur Begräbnisstätte einer Reihe von Familien geworden, die (wie die Horns, Kurcks u. a.) bedeutenden Anteil an der Geschichte Finnlands haben. Der große Brand von Turku 1827 richtete auch am Dom große Schäden an.

Das Bild der mittelalterlichen Kunst in Finnland zeugt auf Grund des Wenigen, was Kriege, Feuersbrunst und Mißachtung übrig gelassen haben, doch von einer starken Selbständigkeit. Sie ist eine rein kirchliche Kunst (die wenigen Burgbauten haben keine kunstwissenschaftliche Bedeutung) und sie entwickelt sich langsam aber stetig mit der Christianisierung des Landes bis zur Reformation, die zunächst das Weiterwachsen der kirchlichen Kunstpflege, welche während 350 Jahren Finnland mit der europäischen Kunst verbunden hatte, schroff unterbrach.

NACH DER REFORMATION

Die Reformation, die in Finnland ruhig verlief, trotz des scharfen Vorgehens des schwedischen Königs und finnischen Oberherrn Gustav Wasa gegen die Kirche, besonders durch die Einziehung kirchlicher Geräte und Kunstschätze, fand in den gebildeten Kreisen starken Widerhall und glich vielleicht auch im Volk manchen über Jahrhunderte erhaltenen Widerstand gegen die Macht der Kirche aus.

Der Kirchenbau ruht ein Jahrhundert und als er um 1700 wieder aufzuleben beginnt, beeinflußt durch die weitere Rodung des Waldes und neue Bauernsiedlungen, spielt nicht der Steinbau, der inzwischen durch Schlösser, Burgen und Städte weitergeführt wurde, sondern der Holzbau die führende Rolle: Von 200 evangelischen Kirchen, die von 1700 bis zur Abtretung Finnlands an Rußland (1809) gebaut wurden, sind nur 11 aus Stein.

Der neue Kirchenbau trägt den veränderten Verhältnissen Rechnung, er zeigt nicht mehr die starke Einheitlichkeit der katholischen Zeit. Die Kirchenmacht ist nicht mehr unabhängig, sondern ein Teil der Staatsmacht. Der Auftraggeber ist nicht mehr ein mächtiger, persönlich interessierter Kirchenfürst, sondern der arme Staat oder die arme Gemeinde. Die großen Orte hatten schon ihre Kirche, und bei Neubauten ging der Auftraggeber meist von geringeren Ansprüchen aus: für die schlichten Formen des lutherischen Gottesdienstes genügten kleinere Kulträume und die billige Herstellung durch das bodenständige, leicht zu bearbeitende Baumaterial, das Holz. Aber mit dem Holzbau findet auch die alte, nie abgerissene

Tradition der Zimmermannskunst neue Aufgaben, ja, sie erlebt hier einen Höhepunkt.

Die finnischen Holzkirchen des 17. und 18. Jahrhunderts sind in zwei Typen vertreten, als Langkirchen und als Kreuzkirchen.

Die Langkirche knüpft an den alten rechteckigen Grundriß der Steinkirche und an ihre äußere Form mit dem Steildach an. Die Wände sind in Blockbautechnik mit horizontalen Lagen ausgeführt. Sie sind durchsetzt mit Strebepfeilern aus kurzen, übereinandergeschichteten Balkenquadraten, die in den Innenraum hineinragen und dort in der Längsrichtung Balken tragen, auf denen die Last des hohen Dachstuhles ruht. Riesige Querbalken verankern die Strebepfeiler der Längswände miteinander. Der Innenraum ist von einer meist tonnenförmigen Bretterdecke überwölbt. Das Steildach ist mit Schindeln gedeckt. Ein Musterbeispiel für den Lang-Kirchentyp ist die 1632 gebaute Kirche von Salo.

Die älteren Langkirchen weisen oft noch ein Element des katholischen Kultes auf, die Chorschranke, eine mit Drechselei oder Schnitzerei verzierte Wand, die den Altarraum vom Gemeinderaum abschloß. In Westfinnland haben diese Langkirchen oft einen in den Westteil eingebauten oder ihm vorgelagerten Turm, der jedoch nur selten ein Glockenturm ist. Auch bei der Holzkirche bleibt der Glockenturm, der schon seit dem Mittelalter fast überwiegend vom Kirchenhaus getrennt wurde, neben dem Gotteshaus. Er ist meist in drei sich nach oben verjüngende Teile gegliedert, mit einfachen, sich aus den Konstruktionsteilen ergebenden Mitteln gestaltet, mit Bevorzugung der barocken Formsprache, wie das charakteristische Beispiel von Petäjärvi (Abb. 21a) zeigt. Um 1700 tritt dann besonders in Österbotten eine Spielart der Langkirche auf, die zur Kreuzform hinüberleitet. Die Mitte der Längswände wird durchbrochen und zwei Kreuzarme, kürzer und niedriger als das Langhaus, werden angebaut (Beispiel von Keuruu, Abb. 20a).

Vom zweiten Typ, von der Kreuzkirche, die zuerst in Südwestfinnland auftaucht, sind schon um 1600 Beispiele feststellbar. Der Grundriß zeigt vier symmetrische Kreuzarme. Der Scheitelpunkt der beiden Dächer ist durch einen Dachreiter geschmückt und betont, der manchmal zu einer Laterne (Antrea, Abb. 20b) ausgebildet ist. Oft erhalten wie in Antrea die Kreuzarme und die Winkel der Arme Anbauten.

Die Herkunft und die Entstehung der Kreuzkirchen in Finnland läßt sich nicht einwandfrei nachweisen. Die kreuzförmige Basilika war ja immer im europäischen Kirchenbau vorhanden und hat in Renaissance und Barock weltberühmte Ausprägungen erhalten. Neben der Entwicklung aus der Langkirche mögen vielleicht

Formen des alten nordischen Kult- und Hallenbaus mitgesprochen haben, die noch in der Zimmermannskunst lebendig geblieben waren. Andrerseits war durch diese neue Form mit einfachen Mitteln eine Zusammenfassung von viel Raum unter einem Dach möglich, so daß auch eine Entwicklung rein aus dem Technischen, unterstützt durch die Anforderung des protestantischen Gottesdienstes, die Predigt möglichst vielen gut zu Gehör zu bringen (die Kanzel liegt an der nordöstlichen inneren Ecke), denkbar ist. Diese neue Form bringt also die klare Betonung eines Mittelraumes mit, ohne sich jedoch schon zu einem wirklichen Zentralbau zu entwickeln. Als dann vom Klassizismus her der reine Zentralbau mit Oberlicht, besonders durch C. L. Engel, in den ländlichen Kirchenbau eindringt (Abb. 21b), findet er ein bereits in der volkstümlichen Baukunst hochentwickeltes Gebilde vor, an das er unmittelbar anschließen kann.

Auch diese protestantische Kirchenbauperiode steht im Zeichen der Volksbaukunst. Die Meister, welche die beiden Holzkirchenformen geschaffen und entwickelt haben, sind größtenteils unbekannt. Sie dürften aber in viel höherem Maße, als bei dem mittelalterlichen Steinbau, Finnen gewesen sein, ebenso wie die Maler, die nun das Innere der Dorfkirche mit bunten Schilderungen in barocker Formsprache schmücken. Auch in der protestantischen Zeit reißt die Bildfröhlichkeit der finnischen Kirchen nicht ab, nicht nur im Reichtum des Schmuckwillens, sondern auch im Bildinhalt, der noch manchen Volksheiligen aus der katholischen Zeit, wie den hl. Henrik und den hl. Georg, die Mutter Gottes und das Kind und andere Elemente des katholischen Kultes weiter ehrt. Es ist besonders die Landschaft Österbotten, wo in Verbindung mit einer starken bäuerlichen Volkskunst im 17. und 18. Jahrhundert auch die Kirchenmalerei ihre lebendige Kraft behält wie etwa in der Kirche von Salo (Abb. 22).

Im 18. Jahrhundert beginnt dann die Abkehr von der ornamentalen Monumentalmalerei. Im Dienste des herben lutherischen Kultes entwickelt sich nur die Tafelmalerei weiter, die sich vom Votivbild zum Altarbild wendet, auf das sich nach und nach der kirchliche Schmuck konzentriert. In der weltlichen Malerei gewinnt das Porträt und die Ausgestaltung anspruchsvoller Wohnräume mit der Steigerung adliger und bürgerlicher Wohnkultur (Abb. 27a) an Boden. Das geschieht in Verbindung mit der Ausbreitung des zünftigen Malerhandwerks, dessen Mitglieder als Illuministen und Konterfeier, Wappenmaler und Miniaturisten das schmuckfreudige Rokoko der gustavianischen Zeit in Finnland vertreten, wie durch den Einfluß der Akademie in Turku (gegründet 1640), die auch die schönen Künste pflegt.

Aus der Gruppe der österbottnischen bäuerlichen Volkskünstler hebt sich eine Persönlichkeit hervor: Mikael Toppelius (1734—1821). Er wirkte in Stockholm an dekorativen Arbeiten beim Bau des Schlosses mit und bestand dort seine Lehrjahre. Dann schmückt er im Laufe seines langen Lebens etwa 40 Kirchen seiner Heimat aus. Von charakteristischen Arbeiten sind die in Paltamo, Pulkkila, Siikajoki, Lochtaja (Lochteå) und Haukipudas zu nennen (Abb. 23). Die Formsprache des Barock und des Rokoko, die er in Stockholm kennen lernte, haben ihm den großen malerischen Schwung gegeben, seine allegorischen Darstellungen, oft im Stil der perspektivischen Deckenmalerei, aber stark dekorativ, bilderbogenhaft, sind urwüchsig und phantasievoll, humorvoll und dramatisch mit kräftiger Charakterisierung seiner oft realistischen Gestalten, in denen er gern Volkstypen porträtiert. Er ist ein Volkskünstler von größter Begabung, beinahe noch anonym und doch schon der vom Anonymen losgelöste freie Typ des Künstlers. Mit ihm schließt wie mit einem letzten großen Akkord die heimatliche monumentale Kirchenmalerei und mit ihm beginnt zugleich die moderne Künstlergeschichte Finnlands im Rahmen der europäischen Kunst.

Die weltliche Baukunst in Finnland bietet bis zu der Neuanlage von Helsinki nur wenig, das von allgemeiner architektonischer Bedeutung wäre. Städtebaulich hatte seit dem Einzug der schwedischen Kreuzfahrerheere nur die alte Hauptstadt Turku Bedeutung bekommen und behalten. Die wichtigsten weltlichen Bauten waren Burgen, militärische Stützpunkte von gewaltigen Ausmaßen, welche die Verwaltung, Beherrschung und Sicherung des Landes gegen den alten Feind im Osten, die Russen, ermöglichen sollten. Die bemerkenswertesten mittelalterlichen Hauptburgen (als Wasserburgen angelegt) stehen in Turku (Abb. 16), Hämeenlinna (Tavastehus), angeblich von Birger Jarl 1249 gebaut, Wiipuri (Wiborg), die große Seefestung gegen Rußland, 1293 von Torkel Knutson errichtet (Abb. 18), und Savonlinna (Nyslott), 1475 von Erik Axelsson Tott erbaut (Abb. 17).

Das Schloß zu Turku wurde um 1300 angelegt. Es bestand ursprünglich aus dem heutigen Westteil, dem alten rechteckigen Burghausbau mit den zwei viereckigen Türmen. Im Laufe der Zeit wurde es der neueren Festungstechnik entsprechend erweitert. Von architektonischer Bedeutung wurde der Umbau und die Innengestaltung, die es durch Herzog Johann erfuhr, dem sein Vater Gustaf Wasa 1556 Finnland als Herzogtum verlieh. Der junge Herzog stattete seinen Hof in Turku fürstlich aus und führte eine glanzvolle Hofhaltung ein, die mit seiner Gefangennahme und Absetzung durch seinen Bruder, König Erik, 1563 endete. Ein Brand im Jahre 1614 vernichtete dieses reiche Denkmal nordischer Renaissance in Finnland. Das Schloß wurde später als Arsenal und Gefängnis benutzt. 1885 wurde darin das Historische Museum der Stadt Turku eingerichtet.

Die kriegerische Epoche der baufreudigen schwedischen Wasakönige brachte den finnischen Adel, die Offiziere und Verwaltungsbeamten stärker mit europäischen Kulturkreisen zusammen. Die höheren Ansprüche an Wohnen und Bauen nach dem Vorbild des königlichen Hofes in Stockholm lassen auch in Finnland, besonders um die alte Hauptstadt Turku herum im sogenannten Eigentlichen Finnland, ländliche Herrenhäuser entstehen, von denen die Schlösser in Qvidja, in Kankas (Abb. 24b) und in Villnäs (Abb. 24a) Beispiele der ersten weltlichen Architektur sind, die in enger Anlehnung an die Adelshöfe in Schweden den Einfluß deutscher und holländischer Renaissance zeigen. Nach der Zeit des großen Unfriedens (1717—1721) und der Russeninvasion Peters des Großen begann eine neue Erholung. Der nun auch von reichen Handelsherren in Auftrag gegebene Herrenhaustyp ist im Typ einfacher und ländlicher. Er ist wie in den gleichzeitigen schwedischen Landsitzen meist ein Hauptgebäude mit zwei einen Hof flankierenden Flügelbauten. Die Architektur zeigt bei den gewöhnlich zweigeschossigen Häusern spätbarocke, einfache symmetrische Aufteilung der Wandflächen durch Fenster und Türen, sparsame Profilierung, ein mehrfach gebrochenes Dach mit zwei symmetrisch angeordneten Schornsteinen, wie das Beispiel Tykö in Perniö (Bjernå), erbaut 1770 von dem in Turku tätigen deutschen Architekten Chr. Fr. Schröder (Abb. 25a).

Unter dem Einfluß des gustavianischen Rokoko dringt der dreieckige Giebel auf der Langseite als konventioneller Bauteil durch, wie der stumpfe Giebel auf der Kurzseite, eine Bauform, die mit Säulen, Pfeilerstellungen und Flügelvarianten vielfach in Holz ausgeführt, im Landhausbau um 1800 zu der einheitlichen Note des Klassizismus (wie das schöne Beispiel von Svartå, Abb. 25b) gelangt.

SEIT 1800

Nach einer siebenhundertjährigen Verbundenheit mit Schweden, während welcher die „finnische Nation" immer als selbständiger Gebietsteil hervorgehoben wurde und sich unter der demokratischen und antifeudalen Verwaltung der schwedischen Regentschaft immer als solcher fühlen konnte, wurde im Frieden von Frederikshamn 1809 Finnland von Schweden abgetrennt und dem russischen Reich angegliedert. Der liberale Kaiser Alexander I. versprach eine weitgehende Selbstverwaltung und garantierte die Privilegien aus der schwedischen Zeit. Aber auch hier begann wie überall in Europa nach Beendigung der napoleonischen Kriege die Einschränkung der demokratischen Errungenschaften. Zum ersten Male lernte Finnland, dessen Geschichte (wie nur bei wenigen europäischen Nationen) immer eine Volksgeschichte war und nie eine dynastische Geschichte, die Schärfe des feudalen Regimentes kennen. Nach wenigen Jahrzehnten begann der Zarismus mit aller Strenge, mit Zensur, Verbannung, militärischer und bürokratischer Bedrückung die Russifizierung des Landes einzuleiten.

Der Widerstand des finnischen Volkes richtete sich im Bewußtsein, daß der Kampf um seine geistige Existenz ging, auch mit geistigen Waffen gegen die Unterdrückung. Es ist gewiß selten, daß der nationale Unabhängigkeitskampf eines Volkes in so hohem Maße von Dichtern und Gelehrten geführt wurde, wie das nun in Finnland geschah. Seine elementare geistige Kraft erhielt dieser Kampf durch eine Volkstumsbewegung. Seit Rousseau und besonders durch die Einwirkung Herders hatte überall in Europa ein Suchen und Forschen nach dem Wesen und der Geschichte der Völker eingesetzt. In der ebenfalls von Herder begründeten Sprachforschung fand diese neue Wissenschaft erste literarische und begriffliche Grundlagen für Volkstum, Volksgeschichte und Volkskunde und erste Maßstäbe für die Bedeutung und Gestaltung nationaler Gemeinschaftswerte und Gemeinschaftsformen. Sie erhielt ihren Auftrieb aus der geschichtlichen Erschließung der mittelalterlichen und frühmittelalterlichen Welt und (im Gegensatz zu der klassischen Bildungswelt) aus der gemanischen Vorzeit, aus ihren Mythen, Götter- und Heldensagen, aus der Entdeckung der großen Völkerfamilien, der Volkslieder, Sagen und Märchen, aus dem Bewußtsein eines Jahrtausende alten kulturellen Erbgutes, das nicht an eine Dynastie oder an einen Stand, sondern an das Volkstum selbst gebunden war. Mit der Entdeckung der Volkstumswerte begann auch in Finnland die große Zeit vaterländischer Besinnung. Dem unter der russischen Unterdrückung erwachenden Nationalbewußtsein zeigten sich plötzlich alte und großartige Überlieferungen der Volkskultur; besonders die Sammlung der finnischen Volksmythen, die Elias Lönnrot (1802—1883) bei den letzten Volks- und Runensängern aufzeichnete und 1835 unter dem Namen „Kalevala" als geschlossenes Epos herausgab, und seine große 1840 unter dem Titel „Kanteletar" erschienene Sammlung der finnischen Volkslyrik erregten weit über Finnland hinaus gewaltiges Aufsehen. Die Kalevala wurde, ebenso wie die Dichtungen Johan Ludvig Runebergs (1804—1877), der in seinen vierunddreißig Balladen „Fänrik Ståls Sägner" dem finnischen Volk ein neues großartiges Volks- und Nationalepos schenkte, die geistige Voraussetzung zu einer das ganze Volk umfassenden Begeisterung für

Heimat und Volkstum. Es ist bezeichnend, daß die Träger dieser Bewegung zunächst den gebildeten schwedischsprechenden Kreisen entstammten, daß von ihnen mit leidenschaftlicher Energie und Liebe die finnische Volkssprache in ihrer reichen Schönheit und Musikalität rasch zu einem Hauptwerkzeug nationaler Selbstbehauptung entwickelt wurde. Diese mächtige geistige Strömung in den ersten Jahrzehnten des 19. Jahrhunderts ist die Grundlage, auf welcher auch die finnische Kunst erwuchs und aus welcher sie noch heute ihre Hauptelemente und ihre Eigenart erhält. In der Geschichte der neueren finnischen Kunst gibt es keine fürstlichen oder geistigen Patronate, keine Förderung und Anregung durch reiche Städte oder Familien, die zur Verherrlichung ihrer Macht oder zur luxuriösen Ausstattung ihres Lebenszuschnittes Künstler heranzogen und eine künstlerische Tradition schufen. Sondern aus dem Volk heraus, im Rahmen der vaterländischen Bewegung, führte die Erkenntnis des Bildungswertes der Kunst zur Kunstpflege und zum Anschluß an die europäische Kunst, der seit dem Ausgang des Mittelalters fast völlig verloren gegangen war.

Es ist nicht oft der Fall, daß eine nationale Bewegung so aus der Kunst heraus, auf Grund der Entdeckung großer nationaler Dichtungen, entsteht und von ihr getragen wird, wie diese Zeit des finnischen Volksfrühlings. In diesem an materieller Kunst so armen Lande trieb die Kraft der nationalen Idee zunächst die moralischen und dichterischen Kräfte an, die sich dann erst der Förderung der bildenden Kunst zuwandten. Finnland war damals und ist noch heute in hohem Maße ein Land der Menschenwerte und nicht der Kunstwerte.

DIE NEUERE BAUKUNST

Mit dem Anfang des 19. Jahrhunderts beginnt in Finnland die Wiederaufnahme einer starken Bautätigkeit, insbesondere durch die Gründung der neuen Hauptstadt Helsinki (Helsingfors). Während der Zugehörigkeit zu Schweden war die alte Hauptstadt Turku der geistige Mittelpunkt. Nach der Trennung Finnlands von Schweden und der Angliederung an Rußland wurde die neue Abhängigkeit (und doch vielmehr Unabhängigkeit) durch die Errichtung einer neuen Hauptstadt dokumentiert, die nach einem Erlaß des Zaren im Jahre 1812 nach Helsinki (damals eine Stadt von etwa 7000 Einwohnern) verlegt wurde. Hier erhielt Finnland ein neues Zentrum, von dem aus der staatliche und geistige Aufbau der Nation erfolgte. Für die Neugestaltung der Stadt sollte ein großzügiger Bebauungsplan aufgestellt und die architektonische Gestaltung in eine Hand gelegt werden. Diese Aufgabe wurde dem deutschen Architekten Carl Ludwig Engel (1778—1840) übertragen, und damit eine Entscheidung von größter Tragweite für die finnländische Baukunst und Kulturgestaltung getroffen. Engel war geborener Berliner, Schüler Gillys, Studiengenosse und Freund Schinkels. Während der Besetzung Preußens durch Napoleon ging er als Stadtarchitekt 1809 nach Reval und Petersburg. Der 1810 zum Generalgouverneur von Finnland ernannte Graf Steinthal berief ihn 1820 in das Stadtbau-Komitee von Helsinki. 1825 wurde er, wie Schinkel in Preußen, Leiter der obersten Baubehörde des Landes. Engel hat nicht nur der neuen Hauptstadt Finnlands Grundriß und Charakter gegeben, sondern überall im Lande, bei kirchlichen und staatlichen Bauten, seinen Einfluß und seine Baugesinnung geltend gemacht. Er hat damit auch eine seit dem Mittelalter nun wieder einsetzende Bautradition in Finnland so begründet, daß noch die heutige finnische Architektur nach den Jahrzehnten der historischen Stilbaukunst an Engels Vermächtnis (was besonders für Helsinki gilt) einen gesunden und unmittelbaren Anschluß finden konnte. Ähnlich wie bei Schinkel bietet die sachliche und aufrichtige Entwicklung seiner Grundrisse und Baukörper, verbunden mit einer schlichten, aber kräftigen dorischen Formsprache mit wenigen Schmuckelementen grundsätzliche Berührungspunkte mit dem neuen Baustil unserer Zeit. Ähnlich wie Schinkel gestaltet auch Engel die Innenausstattung und die Einzelheiten der Ausschmückung eigenhändig mit größter Sorgfalt, auch er ist Maler und Zeichner, Schriftsteller und Verwaltungsbeamter. Die monumentale städtebauliche Aufgabe fand in ihm einen schöpferischen Geist, der die große Linie der Form ebenso beherrscht, wie den technischen Apparat und die handwerklichen Aufgaben. Er ist eine leider in Deutschland viel zu wenig gewürdigte Begabung ersten Ranges, obwohl er in einem selten anzutreffenden Maße der deutschen und preußischen Baukunst seiner Zeit im Ausland ein großartiges Denkmal gesetzt hat. Engels Arbeit beginnt mit der Bearbeitung des Stadtplans von Helsinki. Er erweitert die vorhandenen Maße des Hauptplatzes (des heutigen Senatsplatzes), auf dem sich die repräsentativen Gebäude erheben sollen (Abb. 32). Den vorliegenden kleinlichen Bebauungsplan ändert er zu der großzügigen Anlage ab, die der Hauptstadt für immer den baulichen Kern verleihen wird. Er bestimmt einen von dem bisherigen Plan „links liegengelassenen" Granitblock als Mittelpunkt der Platzanlage und schafft in der Nikolaikirche mit der monumentalen Treppe eine beherrschende Achse. An der Ostseite des Platzes errichtete er zunächst das Senatsgebäude. Als 1827 beim Brand von Turku auch die Akademie vernichtet wurde, ergab es sich von selbst, die neue Universität in der neuen Hauptstadt aufzubauen. Engel legte die Universität, die 1832 eingeweiht wurde, an die Westseite des Senatsplatzes als Gegenstück zu dem Senatsgebäude an der Ostseite. Es entstehen von seiner Hand die schöne Universitätsbibliothek (Abb. 37), das Präfektenhaus im botanischen Garten, das astronomische Observatorium, das Haus des Generalgouverneurs (heute Auswärtiges Amt), des Militäroberinspekteurs, die Kaserne auf Katajanokka (Skatudden) und Kasarmintori (Kaserntorget), die reizende Holzkirche Vanha Kirkko (Gamla Kyrka), in deren Park das Denkmal für die im Freiheitskrieg 1917 gefallenen deutschen Soldaten (Abb. 41) liegt. Weiterhin 20 Landkirchen in Holzbau, wo Engel bei seinen Zentralbauten an den vorhandenen Kreuzkirchenbau (Abb. 21b) anknüpft, die Rathäuser in Pori (Björneborg) und Kristiina (Kristinestad), Krankenhäuser, Villen und Gutshäuser, die Stadtpläne von Turku (wo er als ersten Bau in Finnland das astronomische Observatorium, die heutige Navigationsschule, erbaute) und Porvoo. Neben dieser fruchtbaren und mannigfaltigen Tätigkeit bildete er eine Anzahl von Architekten für die Baubehörde aus. Sein Hauptwerk ist die Nikolaikirche, deren Pläne er von 1818—1839 bearbeitete, deren Grundstein 1830 zur dreihundertjährigen Feier der Confessio Augustana gelegt wurde. Dieser Bau (Abb. 35), welcher die architektonische Krönung der Gesamtanlage sein sollte, und nicht nur im Namen an die Potsdamer Nikolaikirche Schinkels erinnert, hat ihm auch die stärksten Widerstände gebracht. Dreimal wurden seine Pläne in Petersburg verworfen, die vorgeschlagenen Änderungen abgelehnt, so daß der Kampf um sein Hauptwerk die größte Enttäuschung seines Lebens wurde. Als Engel 1846 starb, war die Kirche halb fertig, sie wurde 1852 eingeweiht, und obwohl sie den monumentalen Willen ihres Schöpfers städtebaulich zur Wirkung bringt, trägt sie doch in ihren Einzelheiten alle Zeichen eines Kampfobjektes, das mit Kompromissen abgeschlossen wurde. Die vier Ecktürme und die zwei Flügelbauten an der Treppe nach dem Platz zu sind Zutaten seines Nachfolgers im Amt E. Lohrmann.

Mit Engels Tätigkeit beginnt seit dem Mittelalter die erste große Willensäußerung heimischer Baukunst und künstlerischer Repräsentation.

Aber seine gewaltige Anstrengung, das Land mit einer einheitlichen Baugesinnung zu füllen, blieb zunächst ohne Bestand. Die nächsten Jahrzehnte nach Engels Tod weisen keine einheimischen Baukünstler auf, erst mit dem Beginn der Stilimitationen treten Architekten wie G. Th. P. Chiewitz (Ritterhaus in Helsinki, 1861), Theodor Hoyer (Ateneum und Hotel Kämp in Helsinki,) und G. Nyström (Staatsarchiv, Helsinki) hervor. Dieser Stilhistorismus hat in Finnland einen gemäßigten und weniger protzigen Charakter als etwa in Berlin. Die Reaktion gegen die geliehenen Bauformen der großen Stilepochen vollzog sich deshalb auch in ruhigerer Weise. Die Entdeckung des Ziegels oder des bodenständigen Baumaterials, des Granits und des Bruchsteins, bringt mit neuen Materialwirkungen einen persönlichen Charakter in die alten Stilformen, der dann von den Architekten Hermann Gesellius, Armas Lindgren (1874—1929) und Eliel Saarinen, geb. 1873, zum ersten Male in eine moderne dekorative Bausprache zusammengefaßt wird: in dem finnischen Pavillon der Pariser Weltausstellung 1900, für den auch Akseli Gallén-Kallela seine auf S. 27 erwähnten Kalevalabilder malte. Diese romantische Bauweise der gemäßigten Stilmischungen hat in einer Reihe von Bauten, besonders im finnischen Nationalmuseum, dem Stadtbild von Helsinki eine charakteristische Prägung verliehen. Die Tätigkeit der drei Architekten *, die ein gemeinsames Büro hatten, und die des gleichgerichteten Lars Sonck, geb. 1870, ist die bauliche Parallele zur nationalen Malerei, ist in ihrer falschen Romantik der erste, wenn auch erfolglose Versuch, einen nationalen Baustil zu schaffen. Dieser „nationale" Stil weicht bald einfacheren dekorativen Formen, besonders in den kräftigen Geschäftshausbauten von Sonck (Börse, 1911, und Magazin am Südhafen, 1913—28), wo schon die kubischen Grundformen des Baukörpers klarer in Erscheinung treten. Eliel Saarinen hat nicht nur hervorragende Architekturleistungen für seine Heimat (so die Bahnhöfe in Helsinki und Wiipuri, Abb. 38 u. 39) durchgeführt, sondern ist ein Architekt von internationaler Wirksamkeit. Projekte wie der Friedenspalast im Haag (1906), das Haus der Chicago Tribune (1924), und auch seine städtebaulichen Planarbeiten für Helsinki und für Canberra, die neue Hauptstadt Australiens, haben ihn weltbekannt gemacht. Seit 1923 ist er in Amerika, wo er für den Multimillionär G. G. Booth eine Akademie für Architektur in Canbrook bei Detroit erbaut.

Nach dem Kriege ist, wie überall, auch in Finnland der neue Formwille in der Baukunst durchgedrungen, die romantischen Stilneigungen verschwinden. Die soziale Note, die Forderungen des Verkehrs und der modernen Technik haben einer einfachen und aufrichtigen Grundrißgestaltung zum Siege verholfen und die Formentwicklung von leeren Äußerlichkeiten befreien helfen. Beim Neubau des Kraftwerkes von Inkeroinen (1923) und des Warenhauses Stockmann (1930) in Helsinki von Sigurd Frosterus, geb. 1876, zeigt sich diese unsentimentale, aus der Zweckbestimmung entwickelte Raumeinteilung, ebenso wie bei den Ver-

---

* Bauten von Lindgren: Geschäftshäuser der Versicherungen Suomi, 1909, und Kalevala, 1912, Brändö Casino, 1913, in Helsinki, das Stadthaus in Hanko (Hangö), 1926, Bauten von Sonck außer den schon genannten: Haus der Hypothekenvereinigung und Berghällskirche in Helsinki, 1908, Kirche in Mariehamn, 1928.

waltungs- und Wohnbauten von Väinö Vähäkallio, geb. 1886 (Bürohaus der Genossenschaft Elanto und Hotel in Koli, Abb. 53) und Alvar Aalto, geb. 1898 (Vereinshaus des Landvolkes in Helsinki). Die Baugestalt der Wohnbauten nähert sich im allgemeinen dem Bautyp der Herrenhäuser des Rokoko oder des einfachen Empire, das ja in der gesamten nordischen Architektur der Gegenwart stark durchklingt. Diese ruhige, auch im Grundriß das Experiment vermeidende Haltung prägt den architektonischen Charakter des modernen Helsinki und der größeren Städte. In Helsinki hat der Leiter des Stadtbauamtes, Gunnar Taucher, geb. 1886, in dieser diskreten Formsprache eine Reihe von vorbildlichen Bauten geschaffen, Arbeiterwohnhäuser (Abb. 42), Krankenhäuser, Schulen und Verwaltungsgebäude. Der Gegensatz zwischen Tradition und Funktion wird auch von der jüngsten Architektengeneration nicht übermäßig betont, sondern zum Ausgleich gebracht. Von charakteristischen Arbeiten seien hier die Pauluskirche (Abb. 46) und das Krematorium in Helsinki von Bertel Liljequist, geb. 1885, und die Wohnhausbauten in Tölö (Stadtteil von Helsinki) von K. N. Borg, geb. 1888, genannt, die Begräbniskapelle Erik Bryggmans, geb. 1891, in Parainen (Pargas) (Abb. 48 u. 49), das Gemeindehaus in Tölö von Hilding Ekelund, geb. 1893, und die Kirche in Jyväskylä von Elsi Borg, geb. 1895 (Abb. 47). Martti Välikangas, geb. 1893, hat interessante Ein- und Mehrfamilienhäuser in der heimischen Holzbautechnik (Gartenstadt Käpylä bei Helsinki) ausgeführt und Wohnblocks in Helsinki (Abb. 43) und Tampere (Tammerfors), Jussi Paatela, geb. 1886, der Leiter des finnischen Architektenbundes, Schulbauten auf dem Lande und das Rote-Kreuz-Krankenhaus (Abb. 44) in der Hauptstadt.

Als wichtigste bauliche Anlage der modernen Hauptstadt erscheint das neue Reichstagsgebäude (1931) von J. S. Sirén, geb. 1890, der auch das Denkmal für die deutschen Soldaten in Helsinki (Abb. 41) geschaffen hat. Schon die Gestaltung des Parlamentsplatzes und ihre für die Zukunft geplante Erweiterung sieht hier gewissermaßen ein zweites Forum neben dem Senatsplatz an der Nikolaikirche vor. Der wuchtige geschlossene Baublock bedient sich in seiner Säulenfront einer etwas konventionellen Fassade, die aber durch ihr Ebenmaß die Stilreminiszenzen in den Hintergrund drängt (Abb. 40).

In der finnischen Baukunst der letzten 30 Jahre hat der Wille zur Selbständigkeit und zur nationalen Eigenart immer den Zusammenklang mit der gesamten europäischen Moderne übertönt. Erst dies letzte Jahrzehnt bringt auch hier über die sozialen Probleme der Gemeinschaftsbauten (Wohnungen, Fabriken und Banken), deren Auftraggeber vielfach das in Finnland hochentwickelte Genossenschaftswesen ist, und über die neuen Baumethoden und Bauwerkstoffe den internationalen funktionalistischen Standard, der weder einen Anschluß an die Landschaft noch an die Landesarchitektur zeigt. Als Beispiele der radikalen Moderne seien hier von Valde Aulanko und Erkki Huttunen das Mühlenwerk in Wiipuri (Abb. 50), Alvar Aalto das Tuberkulosesanatorium in Paimio (Abb. 45), das Haus der Zeitung Turun Sanomat in Turku, und die Sulfatfabrik in Toppila (Abb. 51) erwähnt.

DIE MALEREI

Um die Wende vom 18. zum 19. Jahrhundert war in dem kargen und hart um seine wirtschaftliche Existenz ringenden Lande nur eine bescheidene Handwerkskunst vorhanden. Die begabteren Kräfte fanden meist über die Akademie in Turku, wo ein Zeichenmeister die einzige künstlerische Lehrstätte Finnlands leitete, und nach Ablegung ihres Gesellenstückes vor der Malerzunft den Weg an die Kunstakademie nach Stockholm. Wenn sie zurückkehrten, arbeiteten sie wie Toppelius im Rahmen der heimatlichen Volkskunst. Über die Grenzen seines Landes hinaus wirkte als erster Alexander Lauréus (1783—1823). In Turku geboren, lernt er an der Stockholmer Akademie. Inmitten der klassizistischen Formideale seiner Zeitgenossen wendet er sich dem Studium der Holländer Teniers, Adrian von Ostade und Godfried Schalcken zu, und seine Bilder (Abb. 54) beschäftigen sich hauptsächlich mit den malerischen Reizen des Volkslebens. Seine Reisen, die er später als Mitglied der Schwedischen Kunstakademie mit Hilfe der damit verbundenen Reisestipendien machte, führten ihn nach Dänemark, Deutschland, Frankreich und Italien, wo er 1823 in Rom starb.

Lauréus war der Schrittmacher für den Anschluß der jungen finnischen Kunst an die zeitgenössische europäische Kunst. Mit ihm zusammen ist Gustav Wilhelm Finnberg (1784—1833) zu nennen. Auch er kommt aus dem Handwerk, macht 1805 sein Gesellenstück bei der Malerzunft in Turku und geht an die Akademie nach Stockholm. Seit 1817 wirkt er in Turku ein Jahrzehnt als Porträtist und Kirchenmaler. Nach dem Brand von Turku 1827 geht er wieder nach Stockholm, da ihm die Heimat weder Anerkennung noch Entwicklungsmöglichkeiten bietet.

Diese Künstler gehören also ihrer Ausbildung und ihrer künstlerischen Verknüpfung nach nicht so sehr nach Finnland wie nach Schweden. Der nationalen finnischen Kunst fehlte noch ein festes Zentrum, von dem aus Anregungen und Aufgaben gegeben werden konnten. Die alte Hauptstadt Turku stand zu stark in der Nachbarschaft des mächtigen Stockholm. Damals war die Tradition des Mittelalters längst abgerissen, und der kulturelle Einfluß der gustavianischen Epoche war nur in Turku bemerkbar geworden. Öffentliche Bildungsstätten für die Künste fehlten, Museen und Sammlungen waren nicht vorhanden. Nun einen Mittelpunkt für die Förderung der Kunst zu schaffen, wurde die Aufgabe der bürgerlichen Kreise, die sich als Träger der nationalen Bildung fühlten und aus denen heraus 1846 der finnische Kunstverein entstand, der die Förderung und Pflege der Kunst übernahm. Er errichtete 1848 eine Zeichenschule in Helsinki und nahm 1851 die von der Malerzunft in Turku begründete Zeichenschule in Verwaltung. Im Jahre 1887 erhielt der finnische Kunstverein zusammen mit dem 1874 begründeten Kunstgewerbeverein mit Unterstützung des Staates ein eigenes Haus in Helsinki für seinen Unterricht, seine Sammlungen und seine Kunstausstellungen, das heutige Ateneum.

Der finnische Kunstverein hat das Protektorat über die bildende Kunst Finnlands mit Energie und Geschick durchgeführt. Er wurde zum Mittelpunkt und Träger der nationalen Kunstpflege bis in die Gegenwart. Er ist heute noch der Träger des akademischen Kunstunterrichts in Finnland, das keine staatlichen Anstalten dafür eingerichtet hat.

Mit den ersten Bestrebungen des finnischen Kunstvereins ist der Name des Malers R. W. Ekman (1808—1873) verknüpft. Er war in Uusikaupunki (Nystad) geboren, war bei Finnberg in Turku in der Lehre und hatte wie Lauréus mit Hilfe der Stockholmer Akademie Studienfahrten nach Paris, Rom und München machen können. Im Jahre 1845 ging er als Mitglied der Stockholmer Akademie und königlich schwedischer Hofmaler nach Turku und stellte 1846 seine Werke in Helsinki aus, wo er mit Begeisterung begrüßt wurde. Er wird nicht nur mit Schilderungen aus der nationalen Geschichte und dem Volksleben (Abb. 55a) ein Bahnbrecher der landeseigenen Kunst, sondern er widmet sich mit großer Energie der Förderung der Kunst überhaupt, durch seine Lehrtätigkeit an der von ihm mit dem Vorsteher der Malerzunft in Turku begründeten Zeichenschule. Eine Reihe von finnischen Malern sind bei ihm in die Lehre gegangen.

Zu den ersten Schützlingen des Kunstvereins gehören die Brüder Magnus und Ferdinand v. Wright. Magnus zeichnete für das 1845—1852 erschienene Tafelwerk „Finland, framstäldt i teckningar" eine Folge von Landschaften. Beide Brüder haben einen Ruf als Maler von Vögeln (Abb. 55b), ihre Bedeutung liegt aber in der künstlerischen Erschließung der finnischen Landschaft und in der Schilderung der heimatlichen Naturschönheit. Sie leiten mit Ekman das Kapitel Malerei in der neuen vaterländischen Bildungsbewegung ein.

Werner Holmberg (1830—1860), Schüler Ekmans und der Brüder Wright, entwickelt sich über die topographisch - genrehafte Schilderung hinaus und erfaßt Probleme der zeitgenössischen Malerei. Gleich dem Begründer der norwegischen Malerei, J. L. Dahl, folgt er dem großen Zug zur lyrischen Naturschilderung. Wie Dahl in Dresden seine Welt an der Kunst Kaspar David Friedrichs erschloß, so wird für Holmberg die deutsche Romantik der Düsseldorfer Schule und der dort seit 1841 tätige Norweger Hans Gude der Ausgangspunkt seiner Kunst. In fünf Jahren begeisterten Arbeitens wie in Ahnung seines frühen Todes schafft er eine Reihe von Ölbildern und Aquarellen, von denen die schönen finnischen Sommerlandschaften (Abb. 56) zu den besten Bildern seiner Zeit gehören. Als „Naturalist", wie er sich nennt, fühlt er schon die kommenden Probleme der Freilichtmalerei.

Mit Holmberg hat die junge Malergeneration in Finnland über Düsseldorf einen Anschluß an die gesamtskandinavische Malerei gefunden. Auch die Finnländer, die in der Spur Holmbergs nach Düsseldorf gingen, Hjalmar Munsterhjelm (1840—1905) (Abb. 57a), Berndt Lindholm (1841—1914) (Abb. 57b), die erste und hochbegabte Malerin Finnlands, Fanny Churberg (1845—1892) (s. a. S. 35), Fredrik Ahlstedt (1839—1901) (Abb. 58 b) und Viktor Westerhohn (1860—1919) (Abb. 59 a) bleiben in ihrer nationalen Bedeutung mit dem Thema der Landschaftsschilderung verbunden. Erst die letzten, die in Düsseldorf schon die Überalterung der Schule empfanden, folgten dem Zuge der Skandinavier nach Paris, zu Couture, Courbet, Bonnat, Hébert. Zugleich mit dem malerischen Realismus und der Freiluftmalerei nahmen sie auch das soziale Thema auf, unter ihnen besonders Aukusti Uotila (1858—1886) (Abb. 58a), Adolf von Becker (1831—1909) und Gunnar Berndtson (1854—1895). Gerade Berndtson ist ein interessanter und selbständiger Charakter, der neben seiner eleganten zu Meissonier weisenden Bildtechnik auch als erster groß angelegte Porträts malt. In seinen Gesellschaftsbildern, wie in dem feinen novellistischen Bild „Lied der Braut" (Abb. 60) und dem 1885 gemalten Bilde „Rast während der Jahrmarktsreise" (Abb. 59 b), tut er den ersten Schritt zur realistischen, nicht mehr romantisch-verklärten Schilderung von Finnlands Volksleben, einem Thema, das dann Edelfelt, Gallén-Kallela, Järnefeld, Halonen und Rissanen in so charakteristischer Form aufnehmen. Der Anteil Finnlands an der hiermit abgeschlossenen Epoche der Romantik läßt eine typisch skandinavische Note erkennen: Distanzierung in der Begrenzung der Themen, Herbheit und Selbstbeschränkung in der malerischen Ausdrucksform, ein Festhalten an heimischen Motiven. Keiner dieser

Maler hat europäische Bedeutung erlangt, keiner ist aber auch in völlige Abhängigkeit eines Düsseldorfer oder Pariser Vorbildes aufgegangen.

Albert Edelfelt (1854—1905) ist der erste finnische Maler von wirklich europäischer Bedeutung und zugleich der Begründer und Führer einer neuen kraftvollen Malergeneration in der Heimat. Als Sohn eines aus Schweden eingewanderten (und in der Nähe von Porvoo ansässigen) Architekten nimmt er von Hause ein überdurchschnittliches Maß von Bildung mit auf den Weg nach Antwerpen und Paris. Als Schüler der École des Beaux Arts malt er unter dem Einfluß I. L. Geromes seine bekannten Historienbilder, von denen das heute im Ateneum in Helsinki hängende Bild „Herzog Karl verhöhnt die Leiche Klas Flemings" 1878 in der Heimat großes Aufsehen erregte. Bald aber vertauscht er durch die Bekanntschaft mit Bastien Lepage die Atelierarbeit mit der Freiluftmalerei, in der jedoch er mit betonter Zurückhaltung den konsequenten Impressionisten gegenüber sein Lebenswerk gestaltet. Es entstehen die schönen von der Sonnenhelligkeit seiner Heimat durchfluteten Landschaften, in die er seine Fischer und Bauern mit großer Meisterschaft stellt, und seine Volksschilderungen (Abb. 61, 63). Seine Heimat versteht ihn anfangs nicht und der Kunstverein lehnt die „französische Mode" ab.

Nach einem Aufenthalt in England, stark berührt von der Malerei Millais, der Präraffaeliten und Herkomers, entwickelt er seine Kunst der Menschenschilderung in einer Reihe charaktervoller Bildnisse. Zur Geschichte seines Vaterlandes gibt er noch in seinem letzten Jahrzehnt einen künstlerischen Beitrag. In dieser Zeit der Russifizierungsmaßnahmen illustriert er zwei Werke des Dichters Runeberg, den „Kung Fjalar" und besonders das große nationale Epos „Fänrik Ståls sägner". Die Reproduktion seines Aquarells „Björneborger Marsch" (Abb. 62) ist in unzähligen finnischen Häusern als Zeugnis des Freiheitswillens heute noch zu finden.

Während Edelfelt noch der schwedischen Bildungsschicht angehört, wurzelt Akseli Gallén-Kallela (1865—1931) unmittelbar im finnischen Volksboden, hat alle Züge seiner primitiven Kraft, seiner leidenschaftlichen Selbstversenkung und seines Hanges zum widerspruchsvollen Sinnieren. In Gallén erfüllt sich die Wesensart des finnischen Volkstums zu einer großartigen Leistung nicht nur für die finnische, sondern auch für die skandinavische Kunst. Sie ist, so undefinierbar auch solche Kennzeichen erscheinen mögen, typisch nordisch.

Gallén geht 1884 nach Paris an die Akademie Julian, sein großes Erlebnis wird Bastien-Lepage. Die Bilder aus dieser Zeit zeigen den Abstand von der Konvention der Franzosen und auch von Edelfelts Wohlerzogenheit. Wuchtig und grobschlächtig gestaltet er Figuren aus dem finnischen Volksleben. Diese Bilder, die die Leute in

Finnland im Gegensatz zu Edelfelds Werken brutal und häßlich finden, erringen dagegen auf den Ausstellungen, die er gemeinsam mit Edward Munch 1895 in Berlin und Wien macht, Aufsehen. In Berlin, wo er sich auch der Graphik zuwendet, beginnt für ihn im Verkehr mit Munch, Vigeland, Strindberg, Dehmel eine neue Zeit: das Ringen um das Problem der Kunst an sich, das diese Epoche des literarischen Naturalismus, des Symbolismus, der exotischen Einflüsse, des neuerwachten Kunstgewerbes und des Jugendstiles kennzeichnet, in welchem alle die revolutionären Strömungen zu einem ersten großen Versuche für neue Stilbildung zusammenwirken. Der Kampf um die Einheit der Kunst, um die Welt der künstlerischen Magie, über den l'art pour l'art Gedanken hinaus, wurde auch die große Aufgabe in Galléns Schaffen. Um aus der Vielheit seines künstlerischen Wollens und Könnens die Einheit, die Synthese zu erringen, in der auch die angewandte Kunst ein Wesensbestandteil ist, zieht sich Gallén in die Waldeinsamkeit Tavastlands zurück. Er findet den Stoff und den Inhalt dieser Aufgabe im finnischen Volksepos, in der Kalevala, von der er sagt, daß sie ihm wie persönlich erlebt scheine, und findet ihre Form in einer symbolisch monumentalen Malerei.

Das nationale Thema ist bei ihm nicht politisch intellektuell, sondern nur im Mythischen und Sinnbildlichen ausdrückbar: in einer Formsprache, die der Volksdichtung und der Volkskunst verwandt ist. In sieben Jahren schafft er jene Werke, die wie eine Selbstdarstellung der finnischen Volksmythen wirken, die der Zeit des Jugendstils so völlig angehören und doch eine so persönliche Genialität ausstrahlen: „Sampos Verteidiger", „Joukahainens Rache", beide im Kunstmuseum Turku, „Lemminkäinens Mutter" (Abb. 65), „Der Brudermörder", das Motiv aus einer Volksballade, und „Kullervos Fluch" (Abb. 64), im Ateneum in Helsinki. Auf einer Reise in Italien zum Studium der Freskotechnik, 1898, fesseln ihn die Werke Gozzolis und Lorenzettis am Campo Santo in Pisa. Nach seiner Rückkehr malt er dann für den Musiksaal im alten Studentenhaus das Fresko „Kullervo auf der Kriegsfahrt" (Umschlag des Buches) und für ein Mausoleum in seiner Heimatstadt Pori (Björneborg) eine Reihe von Fresken über das Thema Leben und Tod (Abb. 66), wohl seine reifsten, größten künstlerischen Schöpfungen; ferner Bilder mit Kalevala-Motiven in der Halle des Nationalmuseums in Helsinki, die schon in dem finnischen Pavillon auf der Pariser Weltausstellung 1900 Aufsehen erregt hatten.

Im Ringen um die große Synthese der Kunst hat Gallén alle Techniken studiert, er hat seine große dekorative Begabung nach allen Richtungen erprobt, hat geschnitzt, Metall getrieben, Glasmalereien ausgeführt und gebrannt, Möbel, Gewebe, Stickereien entworfen und hergestellt.

Gallén, der 1931 starb, hat in Deutschland starke Anerkennung gefunden, er ist einer der wenigen finnischen Maler, die in Deutschland entscheidende Anregungen empfangen haben, er gehörte zum Kreis der Brücke und war ordentliches Mitglied der Akademie der Künste in Berlin.

Als Zeit- und Kampfgenossen Galléns und Gestalter heimatlicher Themen auf der Grundlage der naturalistischen Malerei gehören Eero Järnefeld (geb. 1863), Pekka Halonen (geb. 1865) und Juho Rissanen (geb. 1873) zusammen. Järnefeld verdankt seiner Schulung an den Akademien in Petersburg und in Paris, wo er wie Gallén und Halonen an der Akademie Julian war, eine gewissenhafte Bildtechnik. Sonst nur wenig beeinflußt, hat er die Tradition Holmbergs und Edelfelts fortgesetzt in stimmungsvoller, fein kultivierter Landschaftsmalerei. Die Stimmung der Natur herrscht auch dort vor, wo er Menschen darstellt, wie in der so charakteristischen Schilderung der Brennrodung (Abb. 69). 1910 und 1920 malt er zwei Monumentalbilder für den Festsaal der Universität in Helsinki. Mit Pekka Halonen steigt wieder ein großes Talent aus der bäuerlichen Bevölkerung auf. Von der Schule des Kunstvereins in Helsinki geht er nach Paris und wird Schüler von Gauguin, der eben aus Tahiti zurückgekehrt ist. Hier löst sich Halonen von der Bindung des Naturalismus, in Gauguins flächigem Stil mit der Leuchtkraft ungebrochener Farben steht er auf der Seite von Gallén: Wie Gallén findet er im Streben nach einer monumentalen Einheitlichkeit das Thema seiner künstlerischen Form im Volkstum. Während aber Gallén leidenschaftlich die Grenzen der Malerei zu überwinden sucht, konzentriert sich Halonen auf das Malerische; wenn Gallén die Helden aus Mythos und Sage schildert und ihnen die Züge des finnischen Volkstums verleiht, monumentalisiert Halonen die Bauern seiner Heimat. Auch er setzt sich wie Gallén besonders unter dem Einfluß von Puvis de Chavannes, den er 1900 in Paris kennen lernt, mit dem dekorativen Stil der Jahrhundertwende auseinander, um sich dann in seiner eigenen Atmosphäre zu so vollendeten und in großartiger Komposition gereiften Werken, wie das eindrucksvolle Bild „Heimfahrt von der Arbeit" 1908 (Abb. 68) durchzuringen.

Auch der dritte der obengenannten Maler, Juho Rissanen, ist bäuerlicher Herkunft, seine Bilder sind noch mehr wie die Halonens Schilderungen von Bauern und Arbeitern. Aber seine Gestalten sind nicht mehr naturalistisch geschildert, im Milieu dargestellt und von ihm getragen, sondern vom Individuellen losgelöst. So hat er eine durchaus eigene, eindrucksvolle Einfachheit, die ihren Charakter von den ererbten

Elementen der Volkskunst erhält: vom Bilderbogen oder von der bäuerlichen Bildweberei, den finnischen Ryenteppichen (Abb. 71).

Mit der Entwicklung der finnischen Malerei von Ekman bis Rissanen ist gewissermaßen das Ringen um die malerische Form, um die künstlerische Einordnung in die europäische Malerei und um eigenwertigen Inhalt abgeschlossen. Es ist erstaunlich, wie reich und selbständig in diesem halben Jahrhundert die finnische Malerei ihren Weg gesucht und behauptet hat, und wie leidenschaftlich dabei der Wille zur Wahrheit und selbsterkämpften Kunst waltete. Das Verhältnis zur Heimat, zu ihrer Landschaft, ihrer Volksart und ihren Menschen bestimmt das Wesen dieser Malergeneration. Es eröffnet ihr zugleich die gemeinsam europäische Linie der Landschaftsmalerei und der realistischen Darstellung, es verleiht ihr aber auch jene Zurückhaltung und eigene Stellungnahme zu den Modeströmungen. Nicht in abhängiger Ergebenheit, sondern abgestimmt und abgegrenzt, werden sie in die finnische Malerei aufgenommen.

Der Kampf gegen den Naturalismus, das Eindringen impressionistischer und nachimpressionistischer Einflüsse aus den verschiedenen Kunstzentren Europas spaltet die auch rein zahlenmäßig stärkere Generation um 1910 in Gruppen, die mehr oder weniger eigene Wege gehen. Hier ist in erster Linie die Malerin Helena Schjerfbeck (geb. 1862) zu nennen.

Die heute siebzigjährige Malerin gehört zu den besten Kräften der finnischen Kunst. Ihre Reisen führten auch sie nach Paris zu Bastien Lepage und Puvis de Chavannes, nach Italien und England. Die englische Malerei (Whistler), die ihrer spirituellen Begabung entsprach, späterhin der Einfluß Toulouse Lautrecs und die ostasiatische Holzschnittkunst haben diesem nach dem Unstofflichen und nach der Wirkung feiner Farbmelodien suchenden Talent Anregungen gegeben. Doch hat sich ihre Kunst erst in den letzten Jahrzehnten zu einer Einfachheit von typisch nordischer, oft an Munch erinnernder Geistigkeit entwickelt (Abb. 76b).

Hugo Simberg (1873—1917) hat neben kräftigen Gemälden und Bildnissen von altmeisterlicher Feinheit (Abb. 76a) höchst originelle Schilderungen von Tod und Teufel und humorvoll-traurigen Ereignissen (Abb. 78) geschaffen. In einer Reihe von Aquarellen und Radierungen behandelt er diese Themen in volksliedhafter Kargheit, aber voll poetischer Phantasie.

Magnus Enckell (1870—1925) vertritt die stärkste malerische Kraft in der Auseinandersetzung zwischen dem Naturalismus Halonens und der Romantik Galléns. Geschult am italienischen Quattrocento und besonders an der religiösen Malerei Zurbarans, den er 1900 für sich entdeckt, faßt er seine Darstellungen in großen tektonischen Zügen zusammen. Erst nach und nach gelangt er von der strengen Dunkelheit seiner ersten Bilder (Abb. 72) unter dem Eindruck Renoirs, Monets, Césannes, Pissarros zu der reinen Palette. Die lichtdurchstrahlten Körper seiner Fresken (Abb. 73) in der von Lars Sonck gebauten Johanniskirche in Tampere (Tammerfors), an deren Ausschmückung auch Simberg mitwirkte, gehören zu den besten Leistungen zeitgenössischer Kirchenmalerei.

Die finnische Kunstgeschichte pflegt den Beginn der Moderne in der heimatlichen Malerei mit der Zeit um 1910 zu verknüpfen, wo der Impressionismus und die von den herkömmlichen Themen aus Landschaft und Volk losgelöste Malerei mit ihren optischen und tektonischen Problemen die künstlerische Auseinandersetzung leiten. Auch diese neue Zeit bringt der finnischen Malerei keine revolutionären Experimentierer, sondern eine Reihe von starken Begabungen, die in handwerklicher Gründlichkeit ihren Anteil an der Moderne leisten. Es ist bezeichnend, daß der einzige Radikale ein Engländer von Geburt, A. W. Finch (1854—1930) war, ein Pointillist aus Seurats Schule, der als Lehrer an der Malschule des Ateneums in langen Jahren zur Problematik der modernen Malerei wohl erheblich beitrug, aber keinen unmittelbaren Einfluß auf seine Generation gewann.

Persönlichkeiten wie Werner Thomé (geb. 1878), einer der wenigen finnischen Maler, die auch in Deutschland studiert haben (er arbeitete bei Zügel in München), Vilho Sjöström (geb. 1873) und Gabriel Engberg (geb. 1872) folgen der Linie Edelfelds in der Vertiefung einer ungezwungenen malerischen Kultur. Sjöström ist wie Engberg ein hervorragender Landschaftsmaler und ein begabter Porträtist von sachlicher Logik und auch Thomé ist einer der Vorkämpfer der Lichtmalerei in seiner Heimat (Abb. 70). Als ein Meister des Porträts, ein vorzüglicher Zeichner und Karikaturist mit einem kräftigen und eleganten Porträtstil (Abb. 74), ist Antti Favén (geb. 1882) zu nennen.

Die in den achtziger Jahren geborene Malergeneration steht noch zu stark unter dem Einfluß revolutionärer Strömungen, als daß sich schon klare Grenzen aufzeigen lassen. Von diesen Malern sei hier Uuno Alanco (geb. 1878), der Leiter der Zeichenschule in Helsinki, hervorgehoben, dessen schöne, architektonisch aufgebauten, flächenhaft formulierten Landschaften schon die Überleitung zum Expressionismus bilden. Hierher gehört auch Mikko Oinonen (geb. 1883), Juho Mäkelä (geb. 1887) und Yrjö Ollila (1887—1932), der in seinen letzten Bildern (Abb. 85) in die kubistische Allegorie und das unpersönliche Programm der technischen Weltbildmalerei hineinreicht. Im Jahre 1914 schlossen sich sieben finnische Maler,

Enckell, Finch, Thomé, Rissanen, Oinonen, Ollila und Laurén zu einer Kampfgruppe zusammen. Ihr Programm war im wesentlichen die Forderung der reinen Farbe. Zwei Jahre später entstand eine neue Gruppe, welche die Kräfte der expressiven Maler zusammenfassen wollte, die Novembergruppe, unter der Leitung von Tyko Konstantin Sallinen (geb. 1879). Sallinen, ursprünglich ein Handwerker vom Lande, trat überraschend in einer Ausstellung 1912 mit einer Reihe von expressionistischen Gemälden auf. Unter Mißachtung der „malerischen Kultur" und der Details gestaltet er Landschaften oder Menschengruppen mit einer ganz elementaren Bewegung durch Konturen und Farbenmassen, deren Disharmonien er rhythmisch zusammenfaßt (Abb. 80, 81). Auch aus Sallinens Malerei spricht wieder eine urwüchsige Gestaltungsleidenschaft, eine blutmäßig gebundene Volksbegabung, die aus der mittelalterlichen Kirchenmalerei wiedererstanden zu sein scheint. Von seinen Weggenossen in der Novembergruppe stehen ihm zwei begabte Maler, Markus Collin (geb. 1882) und Alvar Cavén (geb. 1886), nahe. Sie haben nicht die Wucht Sallinens und auch nicht seine unliterarische Schärfe. Ihre sozialen Schilderungen und Landschaften (Abb. 82, 83) sind mehr illustrativ als unmittelbar. Ilmari Aalto (geb. 1891) (Abb. 79) und Jalmari Ruokokoski (geb. 1886) (Abb. 77b) sind gleichfalls verwandte, im Gegenständlichen kraftvoll und mit sinnlicher Überlegenheit schaffende Persönlichkeiten, die mit Eero Nelimarkka (geb. 1891) (Abb. 77a), Oskari Paatela (geb. 1888), Eero Snellman (geb. 1890), S. W. Sipilä (Abb. 84) zu den besten Kräften der reifen Malergeneration Finnlands gehören.

DIE PLASTIK

Die finnische Bildhauerei im Mittelalter und in den nachfolgenden Jahrhunderten war eine Holzkunst, es sind nur ganz wenige Beispiele von Steinplastik vorhanden. Da bei der Entleerung der Kirchen von religiösem Schmuckwerk am meisten die Plastik betroffen wurde, die im katholischen Kult der Heiligenverehrung besonders gedient hatte, ist in der nachreformatorischen Zeit die Holzbildhauerei so gut wie nicht vorhanden. Der neue Anschluß Finnlands an die europäische Kunst vor hundert Jahren vollzog sich überwiegend auf dem Gebiete der Malerei. Nächst dem großen Bertel Thorwaldsen war Johann Tobias Sergel der erste Klassizist im Norden. Aus Sergels Schule (Sergels Vater war ein aus Thüringen nach Schweden eingewanderter Deutscher) an der Stockholmer Kunstakademie stammt Erik Cainberg, der für die Aula der Universität in Turku einige Marmorreliefs schuf. Aber die Bildhauerei konnte sich in viel geringerem Maße als die Malerei aus wirtschaftlichen Gründen und aus Mangel an Tradition in Finnland entwickeln. Die wenigen privaten Aufträge gingen ins Ausland, die Skulpturen für die neue Nikolaikirche in Helsinki z. B. wurden 1850 an deutsche Bildhauer (Wredow und Schievelbein) vergeben*.

Die finnische Literaturgesellschaft zog durch den Auftrag für eine Statue Porthans, des Vaters der finnischen Geschichte und Sprachforschung, die in Turku aufgestellt wurde, den schwedischen Bildhauer C. E. Sjöstrand (1828—1906) ins Land. Hier wurde er von der nationalen Bewegung zu Darstellungen aus dem Wirkungskreis der finnischen Literaturgesellschaft angeregt. Durch seine klassizistische Formulierung der finnischen Mythenfiguren und einer Reihe von Porträts aus dem Kreis der nationalen Dichter (Cygnäus, Runeberg, Lönnrot, Topelius) hat er (wie Ekman der Malerei) der finnischen Bildhauerkunst besonders als Lehrer der Plastik den Weg gewiesen. Walter Runeberg (1838—1920), ein Sohn des Dichters Runeberg, der bei Sjöstrand und dem Thorwaldsenschüler Bissen in die Lehre ging, wird ganz von dem Thema der klassischen Mythologie gefangen genommen. Für Helsinki schuf er die Denkmäler seines Vaters Johann Ludwig Runeberg und des Zaren Alexander II. Neben ihm steht Johannes Takanen (1849—1885), auch ein Schüler Sjöstrands, dessen herbere und doch sinnlich ausdrucksvollere Arbeiten eine unmittelbarere, unliterarische Begabung zeigen, deren Entwicklung durch seinen frühen Tod unterbrochen wurde. Der Realismus fand auch in der Plastik einen echten und sehr kräftigen Vertreter in Robert Stigell (1852—1907). Er kommt vom Steinhauerhandwerk her, lernt bei Sjöstrand und in Rom, wo er mit Runeberg und Takanen zusammenarbeitet, findet dann in Paris in Carpeaux und Frémiet die Meister der repräsentativen Plastik des zweiten Kaiserreiches, deren dramatisches Temperament sich in Stigell's bekannter Gruppe „Die Schiffbrüchigen" auf dem Observatoriumsberg in Helsinki äußert. Er ist eine finnische Parallelerscheinung zu dem Norweger Vigeland. Ganz im französischen Geschmack und Geist ist die Tätigkeit Ville Vallgrens (geb. 1855). Seine kapriziöse Geschicklichkeit in der Formulierung von plastischen Ideen hat ihn nur selten, wie in seiner besten Arbeit, dem fröhlichen volkstümlichen Brunnen am Hafenplatz in Helsinki (Abb. 33), zur Vollendung von monumentalen Werken kommen lassen. Seine Kleinplastik, von der ein großer Teil im Vallgrenmuseum in Porvoo vereint ist, bildet einen selten anzutreffenden plastischen Beitrag zum Gestaltungskampf des Jugendstils. Ein

* Die Modelle in der Katharinenkirche in Brandenburg.

Vertreter des konsequenten Realismus ist Emil Wikström (geb. 1864). Seine bekannteste Arbeit ist der Fries im Giebel des Ständehauses der Hauptstadt.

Auch in der Plastik teilt sich die neuere Künstlergeneration in verschiedene Gruppen, es entwickeln sich Talente, die abseits der Konvention eigene Wege gehen, Victor Malmberg (geb. 1867), ein Schüler Stigells mit starken Anklängen an Sinding und Rodin, Emil Halonen (geb. 1875), der als erster wieder Holzplastik macht, gleich seinem Vetter Pekka Halonen zum Stilisieren neigend, der Tierbildhauer Freiherr Emil Cedercreutz (geb. 1879) und Galléns Schüler Alpo Sailo (geb. 1877).

Als ein Plastiker von Rang tritt Felix Nylund (geb. 1879) (Abb. 92a) hervor und Into Saxelin (1883—1929) (Abb. 86) mit meisterhafter Steinarbeit an kleinen Figuren.

Die große eigene Note in der plastischen Kunst Finnlands setzt aber ein, als der bodenständige Werkstoff erschlossen wird: der Granit, in dem auch Nylund schon starke Lösungen gefunden hat. Dies ist nicht nur ein äußeres Moment. Auch dort, wo die beiden Meister der Granitplastik, Johannes Haapasalo (geb. 1880) und Väinö Aaltonen (geb. 1894) in anderem Material, Marmor, Stein oder Bronze arbeiten, übertragen sie darauf doch die Strenge und plastische Spannkraft, die sie dem harten Granit abgerungen haben.

Haapasalos weiblicher Akt im Ateneum zu Helsinki aus rotem Granit (Abb. 87) ist nicht nur in der Behandlung des Materials, sondern auch künstlerisch eine an die ägyptischen Granitbildwerke erinnernde Leistung. In Aaltonen ist der finnischen Kunst ein Meister von Weltrang erstanden. Mit ungewöhnlichem Genie hat er schon als junger Anfänger Werke von vorbildlicher Beherrschung der Einzelheit im Rahmen monumentaler Formen geschaffen. Fern von jedem Stilisieren, von jedem realistischen Effekt, mit unübertrefflicher Spannung läßt er den Eindruck unmittelbar aus dem geformten Raum, aus der Melodie des körperlichen Geheimnisses entstehen. Die Darstellung des Olympiasiegers Nurmi (Abb. 88), schon technisch ein Meisterwerk, ist eine künstlerische Versinnbildlichung der Körperkultur und des Laufens, die in den wenig geglückten Darstellungen des Sports in der modernen Kunst einzig dasteht. Seine weibliche Figur aus schwarzem Granit (Abb. 90) im Ateneum, die finnische Venus, gehört zu den eindrucksvollsten Schöpfungen der plastischen Kunst überhaupt.

Man könnte glauben, daß die plastische Begabung der finnischen Künstler größer sei als die malerische, so stark und überragend treten Bildhauer wie Haapasalo, Aaltonen, Victor Jansson (Abb. 91) und G. Tigerstedt (Abb. 92b) und der Holzbildhauer Hannes Autere (Abb. 93) mit Leistungen hervor, die in noch höherem Maße als die der Malerei im Rahmen der europäischen Kunst gewürdigt zu werden verdienen.

VOLKSKUNST UND KUNSTHANDWERK

Finnland weist, wie die von der technischen Zivilisation langsamer und später erfaßten nord- und osteuropäischen Länder, eine starke bodenständige Volkskunst auf. Sie hängt in der Vergangenheit mit der abgeschlossenen Nationalwirtschaft zusammen, mit dem Zwang zur Selbsterzeugung handwerklicher Güter auf dem weiten Lande, mit der Genügsamkeit im Wohn-, Wirtschafts- und Lebenszuschnitt und mit der geschlossenen bäuerlichen Gemeinschaftskultur. In der Gegenwart ist von dieser bäuerlichen Volkskunst praktisch nicht viel übrig geblieben, ideelich aber ist der in ihr wurzelnde Handwerksgedanke in vorbildlicher Weise (wie in den anderen nordischen Ländern auch) für neues Kunsthandwerk fruchtbar gemacht worden, und er hat, worauf wiederholt hingewiesen wurde, auch auf die hohe Kunst grundsätzlichen Einfluß ausgeübt.

Die bäuerliche Volkskunst hat in Finnland hauptsächlich in Österbotten eine großartige Entfaltung gefunden, besonders, wie überall in den europäischen Volkskunst, in der Zeit des farbenfreudigen Rokoko, das ja auch den finnischen Holzkirchen jene schönen Malereien als Nachklang der mittelalterlicher al-secco-Kunst geschenkt hat. Sehr viel später als etwa in Deutschland hat sich hier die freie Kunst vom Handwerk gelöst, sehr viel länger hat das Handwerk den Volkskunstcharakter bewahrt, das in dem städtearmen Land nicht die berufsständische Rolle wie etwa in Mitteleuropa spielen konnte, sondern im ländlichen Haushandwerk die stärkste Vertretung fand.

In der bäuerlichen Volkskunst sind wie überall so auch hier zunächst die Männer- und die Frauenarbeiten zu unterscheiden, und in den Sachgebieten die Festgaben und das Alltagsgerät. Der überwiegende Anteil der Volkskunst, jedenfalls der anspruchsvolle, kunstvolle, wird ja im Dienste des Brauches hergestellt, es sind Festgaben, meist Minnegaben, die der Bursche für das Mädchen, die Braut für den Bräutigam macht, oder Geräte für den neuen Haushalt. Und es sind meist Gegenstände der Hauswirtschaft, Kästchen und Truhen, Spinn- und Webegeräte, Mangelbretter, Näpfe, Käseformen, Pferdekummete (Abb. 103) auf der einen Seite, oder Gewebe und Stickereien auf der anderen Seite. Das Schnitzmesser und das Holz vertreten die Männerarbeit, Webstuhl und Nadel, Gewebe und Stickereien die Frauenarbeit. Geographisch betrachtet überwiegt in Ostfinnland, in Savolax und Karelien, in der Qualität die textile Frauenarbeit, besonders in Stickereien und Klöppelspitzen, die aber auch an der Westküste in Rauma beheimatet sind, während in Mittel- und Westfinnland auch die Männerarbeit auf beachtenswerter Höhe steht. Die Gestaltung des Ornamentes zeigt an den Hornarbeiten, den Holzschnitzereien und Stickereien das alte prähistorische Ornamentgut geometrischer Art, besonders die nordeuropäische Kerbschnittornamentik, dazu die gegenständlich-sinnbildlichen Motive der Volkskunst (Lebensbaum, Doppeltiere, Kreuz, Rad, Stern, Herz und ihre Abwandlungen), durchsetzt mit den Formelementen der Stilepochen, wobei in den ostfinnischen Ornamenten ein deutlicher Gegensatz zu der westlichen Ornamentik erkennbar ist. Das finnische Nationalmuseum in Helsinki und seine Freiluftabteilung, das Inselmuseum Seura-Saari in den Schären von Helsinki, bieten vorzügliche Sammlungen der finnischen Volkskunst.

Eine besondere Stellung in der finnischen Volkskunst nehmen die Ryen ein. Es sind auf dem Webstuhl hergestellte Bettdecken (meist Brautgaben), Teppiche mit etwa 2—3 cm langen, im Smyrnaknoten eingeknüpften Noppen, die der Oberfläche ein fellartiges Aussehen verleihen. Das Wort rya (finnisch ryijy) kommt aus dem Altnordischen und hängt mit dem Worte ruh-rauh, zottig, zusammen, dessen Wurzel auch in dem deutschen Wort „Rauchwaren" erkennbar ist. Diese Noppenwebereien scheinen außer in Finnland und Schweden im ganzen Ostseegebiet vorhanden gewesen zu sein. Auch in Schleswig-Holstein gibt es Kissen und Decken in Noppen-Technik und in Ospreußen Teppich-Decken, die den finnischen nahestehen. Innerhalb der skandinavischen Ryen zeichnen sich die finnischen auf Grund der alten Pflanzenfärbekunst durch große Farbenschönheit und starke bildliche Eigenart aus. Die finnischen Frauen haben mit den Ryen nicht nur für die finnische Volkskunst, sondern auch für die Teppichweberei ganz allgemein einen Beitrag geleistet, der zu den eindrucksvollsten textilen Leistungen überhaupt gehört (Abb. 94, 95). Sie bilden mit den Kalevalagesängen die schönsten Zeugnisse der finnischen Volkskultur.

Die Wiederbelebung der alten volkstümlichen Handwerkskunst ist in den skandinavischen Ländern auch die Grundlage für das neuzeitliche Kunsthandwerk gewesen. Sie ging in Finnland von der Tätigkeit der Malerin Fanny Churberg aus, die 1879 den Verein der Handarbeitsfreunde (Suomen Käsytion Ystävät, Helsinki) gründete. Dieser Verein hat nicht nur eine hochstehende moderne Textilkunst in Finnland ins Leben gerufen, in der besonders die Ryenweberei wieder Hervorragendes leistet (Abb. 96, 97), sondern hat auch durch die Einwirkung auf den Handarbeitsunterricht in den Schulen und den Volksbildungsanstalten das Verständnis für den Wert der textilen Frauenarbeit wieder in weite Kreise getragen. Das allgemeine Interesse für Kunsthandwerk wurde durch den Einfluß des Jugendstils gefördert, der neue individuelle Bedürfnisse und Anregungen für die Innenausstattung und für die dekorative Belebung des Baus vermittelte. Auch hier waren es Maler und Architekten, die selbst Kunsthandwerker wurden oder für Möbelbau, Keramik und Glasarbeiten, Textilien und Metallarbeiten die neuen Wege wiesen, wie die Architekten Saarinen, Lindgren und Gesellius. Der Anteil von Gallén-Kallela an der dekorativen Kunst Finnlands wurde schon erwähnt, bei ihm ist der Zusammenhang mit der Volkskunst besonders deutlich, nicht so sehr bei der Tätigkeit des Malers A. W. Finch, der als erster die moderne Keramik im Sinne van de Veldes in der Werkstätte Iris in Porvoo und als Lehrer an der Kunstgewerbeschule in Helsinki einführte. Die Abkehr vom Individualismus des Jugendstils hat auch in Finnland zu einem ruhigen handwerklichen Kunstgewerbe geführt, das auf textilem Gebiet, aber auch in Metall- und Eisenarbeiten, im Möbelbau und in der Glaskunst (Abb. 104) ganz vorzügliche Leistungen aufweist, die in ihrer Gesamtstimmung an das nordische Empire anklingen und ohne schematischen Radikalismus sich der neuzeitlichen Bau- und Wohnkultur einfügen.

LITERATUR

Kalevala, das Volksepos der Finnen, deutsch von Hermann Paul, Helsinki 1885
Kanteletar, die Volkslyrik der Finnen, deutsch von Hermann Paul, Helsinki 1882
Kaarle Krohn: Kalevalastudien, Helsinki 1925
Johannes Öhquist: Das Löwenbanner. Des finnischen Volkes Aufstieg zur Freiheit. Berlin 1923
Johannes Öhquist: Finnland, Leipzig und Berlin 1919
Johannes Öhquist: Finnland, Berlin 1928
F. Thierfelder und J. Öhquist: Suomi-Finnland, Berlin 1924 und 1932
Graf Rüdiger von der Goltz: Meine Sendung in Finnland und im Baltikum, Leipzig 1920
Joh. Öhquist: Neuere bildende Kunst in Finnland, Helsinki 1930
J. Tikkanen: Die moderne bildende Kunst in Finnland, Helsinki 1925
L. Wennervirta: Finnlands Kunst, Helsinki 1926 (schwedisch)
L.Wennervirta: Goottilaista monumentaalimaalausta länsi-suomen ja ahvenanmaan kirkoissa (Gotische Monumentalmalerei in den Kirchen Westfinnlands u. Ålands). Ausführl. Anhang in Deutsch, Helsinki 1930
Onni Okkonen: L'art finlandais aux XIXe et XXe siècles, Helsinki 1932 (französisch)
Baukunst in Finnland. Herausg. v. Finnlands Architektenverband. Helsinki 1932 (finnisch, schwed., engl.)
U. T. Sirelius: Die finnischen Ryen, Helsinki 1926 (englisch)

Nur in finnischer Sprache erschienene Veröffentlichungen sind für den deutschen Leser nicht erwähnt.

DIE WICHTIGSTEN MUSEEN FINNLANDS

Helsinki (Helsingfors): Finnisches Nationalmuseum (Kulturgeschichte, Vorgeschichte, Volkskunde, Volkskunst) und Freilichtmuseum. Ateneum (Malerei u. Plastik). Kunstgewerbemuseum. Eisenbahnmuseum. Stadtmuseum. Zoologisches Museum.
Turku (Åbo): Historisches Museum der Stadt Turku im alten Schloß. Kunstmuseum.
Vaasa: Historisches Museum von Österbotten (Volkskunst aus Österbotten und Geschichte) nebst den historischen und Kunstsammlungen von Prof. Karl Hedman.
Tampere (Tammerfors): Museum von Häme (Volkskunst aus Häme [Tavastland], mittelalterliche kirchl. Kunst u. a.). Kunstmuseum.
Pori (Björneborg): Museum von Satakunta (hauptsächl. Volkskunst aus Satakunta).
Viipuri (Wiborg): Museum der Stadt Viipuri (Geschichte, Volkskunst). Kunstmuseum.
Kuopio: Museum der Vaterländischen Gesellschaft zu Kuopio (Volkskunde, aus Savo und Karelien, zoologische und botanische Sammlungen).

Von den abgebildeten Kunstwerken befinden sich im Finnischen Nationalmuseum, Helsinki: 8—15, 27a, 94, 95, 98—103; im Ateneum, Helsinki: 54—66, 69—72, 75—82, 87, 88, 90—92; in der Sammlung Keirkner, Helsinki: 67, 73; in der Sammlung Hedman, Vaasa: 68; im Finnischen Reichstag, Helsinki: 74.

BILDVORLAGEN

Finnisches Nationalmuseum Helsinki: 1, 4, 6 bis 15b, 20a, 21 bis 23, 26b, 27a, 38, 94, 95, 98 bis 103. Historisches Museum Turku: 27b. Dr. Kurt Antell, Helsinki: 24 und 25. Prof. Jussi Paatela, Helsinki: 29 und 31. Marie Jaedicke, Berlin: 41. Dr. Wennervirta, Helsinki: 84. P. J. Bögelund, Helsinki: 35. Foto Kaleva, Helsinki: 51. Foto Commercial, Helsinki: 54 bis 67, 69 bis 73, 75 bis 85, 87 bis 93. Foto Iffland, Helsinki: 5, 18, 28b, 39, 40, 46, 53, 74. Foto Pietinen, Helsinki: 30a, 3, 15a, 37, 45. Sundström, Helsinki: 25a, 97. Foto Kansan Kuvalehti, Porvoo: 26a. Foto Heurlin, Turku: 2. Fotomagazin Solio in Vaasa: 68. Finnische Luftfahrt: 32. Abb.22 nach einem Aquarell von Prof. Armas Lindgren.

Es lieferten: das Papier: Scheufelen, Oberlenningen, den Druck des Textes: Otto v. Holten, den Druck der Bilder: Felgentreff & Co., die Druckstöcke: Bendix & Lemke, den Einband: Biblos, alle in Berlin

DIE BILDER

1. GRAVIERTE BRONZEPLATTE VOM ST. HENRIKSGRABE IN NOUSIAINEN (NOUSIS). TEXT SEITE 8

2. DOMKIRCHE ZU TURKU (ÅBO) 13.—15. JAHRH. TEXT S. 10

3. MITTELSCHIFF DER DOMKIRCHE ZU TURKU (ÅBO). TEXT S. 11

1. KIRCHE IN SAUSO (SAGU) 13. JAHRH. TEXT S. 8

7. INNENANSICHT DER KIRCHE VON LOHJA (LOJO) 15. JAHRH. TEXT S. 9

6. ALTARSCHREIN AUS DER KIRCHE IN URJALA (URDIALA) ANFANG 13. JAHRH.   TEXT S. 10

9. ALTARSCHREIN MIT ST. ERIK UND ST. HENRIK. SCHWEDISCHE ARBEIT. ENDE 15. JAHRH.

10. MADONNA AUS DER KIRCHE VON NOUSIAINEN (NOUSIS), 11. JAHRH.
GOTLÄNDISCHE ARBEIT

11 a. MADONNA AUS DER KIRCHE VON PERNAJA (PERNÅ) EINHEIMISCHE ARBEIT. MITTE 14. JAHRH. TEXT S. 10

11 b. DARSTELLUNG DER HL. MARGARETA AUS DER KIRCHE VON VEHMAA (VEHMO) DEUTSCHE ARBEIT DES 15. JAHRH.

12. TAFELBILD DES ALTARSCHREINS AUS DER KIRCHE VON UUSIKIRKKO (NYKYRKA) MIT DER DARSTELLUNG DER BARBARA-LEGENDE VOM MEISTER FRANCKE AUS HAMBURG, 1425—1450. TEXT S. 10

13. TAFELBILD DES ALTARSCHREINS AUS DER KIRCHE VON UUSIKIRKKO (NYKYRKA) MIT DER DARSTELLUNG DER BARBARA-LEGENDE VOM MEISTER FRANCKE AUS HAMBURG. 1425—1430. TEXT S. 10

14. CHORGESTÜHL AUS DER KIRCHE VON HOLLOLA MIT DEM BILDNIS OLAF DES HEILIGEN UND DEM WAPPEN DES BISCHOFS MAGNUS STJERNKORS. SPÄTES 15. JAHRH. TEXT S. 10

15. MITTELALTERLICHE SCHNITZEREIEN AUS DER KIRCHE VON HOLLOLA. TEXT S. 10

16. DAS SCHLOSS ZU TURKU (ÅBO), ERBAUT UM 1300. TEXT S. 11

17. DIE OLOFSBURG IN SAVONLINNA (NYSLOTT), ERBAUT 1475. TEXT S. 14

18. DIE BURG IN WIIPURI (WIBORG), ERBAUT 1293. TEXT S. 14

19. KIRCHE IN NAANTALI (NÅDENDAL)

20 a. HOLZKIRCHE IN KEURUU MIT ANGEBAUTEN KREUZARMEN, 1758. TEXT S. 12
20 b. HOLZKIRCHE IN ANTREA, KREUZKIRCHE MIT ANBAUTEN. TEXT S. 12

21 a. KREUZKIRCHE IN PETÄJÄRVI MIT GLOCKENTURM. TEXT S. 12
21 b. ZUM ZENTRALBAU ENTWICKELTE KREUZKIRCHE IN PYHÄJOKI VON C. L. ENGEL. TEXT S. 13, 19

22. INNERES DER HOLZKIRCHE VON SALO IN ÖSTERBOTTEN, 1632. TEXT S. 13

23. HIOBS SPEISUNG, MALEREI VON MIKAEL TOPPELIUS, 1774
IN DER KIRCHE VON HAUKIPUDAS. TEXT S. 14

24 a. HERRENHAUS IN VILLNÄS, ERBAUT UM 1650. TEXT S. 15
24 b. HERRENHAUS IN KANKAS, ERBAUT UM 1550. TEXT S. 15

25 a. HERRENHAUS IN TYKÖ, ERBAUT 1766. TEXT S. 15
25 b. HERRENHAUS IN SWARTÅ, ERBAUT 1783. TEXT S. 15

26 a. WAPPENSCHILDER DER FAMILIEN BLÅFIELD UND VON DER LINDE, MITTE 17. JAHRH.
26 b. WOHNRAUM DES HERRENHAUSES IN VILLNÄS

27 a. WOHNRAUM DES HERRENHAUSES VON JACKARBY. 1760. TEXT S. 13
27 b. EMPIREZIMMER AUS DEM HISTORISCHEN MUSEUM IN TURKU (ÅBO)

28 a. BAUERNHOF AUS HAUHO IN HÄME (TAVASTLAND), WESTFINNLAND
28 b. BAUERNSTUBE AUS HARTOLA, MITTELFINNLAND

29 a. BAUERNHOF AUS SAVO (SAVOLAKS) MITTELFINNLAND
29 b. BAUERNHÖFE AUS LAPPO IN POHJANMAA (ÖSTERBOTTEN)

30 a. KLEINSTBAUERNHOF AUS HÄME (TAVASTLAND)
30 b. FINNISCHE BÄUERLICHE BADESTUBE

31 a. SPEICHER AUS SÜDFINNLAND
31 b. ALTER SPEICHER AUS HÄME (TAVASTLAND)

32. LUFTBILD DES SENATSPLATZES VON HELSINKI (HELSINGFORS) MIT DER NIKOLAI-KIRCHE. TEXT S. 13

33. VILLE VALLGREN, BRUNNEN AM HAFENPLATZ IN HELSINKI

34. HAFENBILD VON HELSINKI

35. NIKOLAIKIRCHE IN HELSINKI, ERBAUT NACH DEN PLÄNEN VON C. L. ENGEL. 1830—1852. TEXT S. 19

36. STADTBIBLIOTHEK IN TURKI (ÅBO)

35. UNIVERSITÄTSBIBLIOTHEK IN HELSINKI, ERBAUT VON C. L. ENGEL. TEXT S. 19

38. ELIEL SAARINEN. HAUPTBAHNHOF IN HELSINKI. TEXT S. 29

39. ELIEL SAARINEN, BAHNHOFSGEBÄUDE IN WIIPURI (WIBORG)   TEXT S. 20

J. S. SIREN, REICHSTAGSGEBÄUDE IN HELSINKI, 1931. — TEXT S. 31

31. DENKMAL IN HELSINKI FÜR DIE IM FINNISCHEN UNABHÄNGIGKEITSKAMPF 1917—1918 GEFALLENEN DEUTSCHEN SOLDATEN VON J. S. SIRÉN. TEXT S. 19, 21

12. GUNNAR TAUCHER, ARBEITERWOHNBLOCKS IN HELSINKI. 1925. TEXT S. 21

63. MARTTI VÄLIKANGAS, WOHNHAUSBLOCK IN HELSINKI, 1926. TEXT S. 21

11. JUSSI PAATELA, ROTES-KREUZ-KRANKENHAUS IN HELSINKI, 1932. TEXT S. 21

45. ALVAR AALTO, TUBERKULOSE-SANATORIUM IN PAIMIO, 1932. TEXT S. 22

16. BERTEL LILJEQUIST, PAULUSKIRCHE IN HELSINKI 1931. TEXT S. 21

17. ELSIE BORG, KIRCHE IN JYVÄSKYLÄ. TEXT S. 21

18. ERIK BRYGGMAN: BEGRÄBNISKAPELLE IN PARGAS 1930. TEXT S. 21

49. ERIK BRYGGMAN, INNERES DER BEGRÄBNISKAPELLE IN PARGAS. 1930. TEXT S. 21

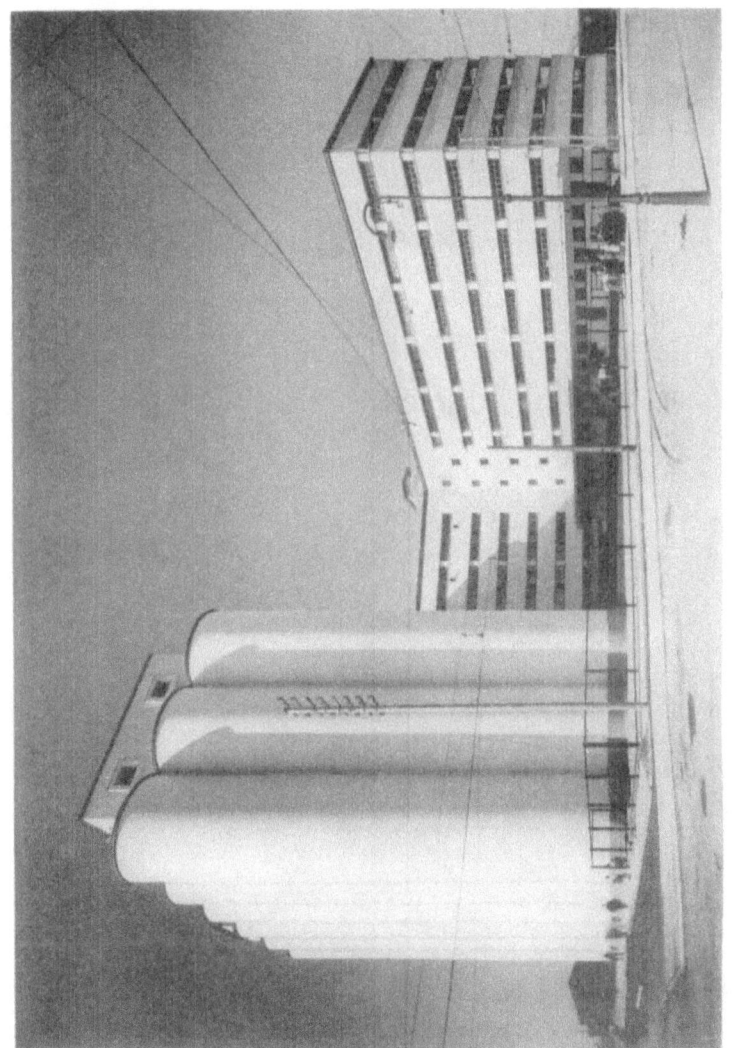

38. AARNE AALTO UND ERKKI HUTTUNEN, MÜHLENWERK IN WIIPURI (VIBORG), 1935. TEXT S. 22

51. ALVAR AALTO. SULFATFABRIK IN TOPPILA. 1931. TENT 8. 22

22. ARBEITERWOHNUNGEN IN KARHULA BEI KOTKA

33. VÄINÖ VAHAKALLIO. TOURISTENHOTEL IN KOLI. 1930. TEXT S. 21

54. ALEXANDER LAURÉUS, GEIGENSPIELER, 1809. TEXT S. 22

55 a. R. W. EKMAN, SZENE AUS RUNEBERGS ELENJÄGER, 1856. TEXT S. 23
55 b. F. v. WRIGHT, ELSTERN, 1867. TEXT S. 23

56 a. WERNER HOLMBERG, LANDSCHAFT AUS BIRKALA, 1858. TEXT S. 24
56 b. WERNER HOLMBERG, STURM AUF DEM NÄSIJÄRVI, 1859. TEXT S. 24

57 a. HJALMAR MUNSTERHJELM, WEG IN TAVASTLAND, 1883. TEXT S. 24
57 b. BERNDT LINDHOLM, STRAND MIT FISCHERN, 1885. TEXT S. 24

58 a. AUKUSTI UOTILA, HERBSTLANDSCHAFT IM MONDSCHEIN, 1883. TEXT S. 24
58 b. FREDRIK AHLSTEDT, AUSSICHT VOM AURAJÄRVI, 1872. TEXT S. 24

59 a. VICTOR WESTERHOLM, AUSSICHT AUF ÅLAND, 1896. TEXT S. 24
59 b. GUNNAR BERNDTSON, RAST AUF DER JAHRMARKTSREISE, 1886. TEXT S. 24

66 GUNNAR BERNDTSON, DAS LIED DER BRAUT, 1881. TEXT S. 24

61. ALBERT EDELFELT, FRAUEN AN DER KIRCHE VON RUOKOLAHTI, 1887. TEXT S. 25

62. ALBERT EDELFELT, DER BJÖRNEBORGER MARSCH, 1892. ILLUSTRATION ZU RUNEBERGS „FÄNRIK STÅLS SÄGNER". TEXT S. 25

63. ALBERT EDELFELT, EINES KINDES LETZTE FAHRT, 1879. TEXT S. 25

64. AKSELI GALLÉN-KALLELA, KULLERVOS FLUCH, 1899. TEXT S. 26

65. AKSELI GALLÉN KALLELA, LEMMINKÄINENS MUTTER, 1897. TEXT S. 76

66. AKSELI GALLÉN-KALLELA, STUDIE FÜR FRESKOMALEREIEN IN EINEM MAUSOLEUM ZU BJÖRNEBORG, 1903. TEXT S. 26

67. AKSELI GALLEN-KALLELA, BILDNIS MAXIM GORKI, 1906

68. PEKKA HALONEN, HEIMFAHRT VON DER ARBEIT, 1905. TEXT S. 27

69. EERO JÄRNEFELT. BAUERN BEIM WALDRODEN. 1893. TEXT S. 27

20. WERNER THOMY, SPIELENDE KNABEN. TEXT S. 29

71. JUHO RISSANEN. DIE WAHRSAGERIN. AQUARELL. 1899. TEXT S. 28

72. MAGNUS ENCKELL, GETHSEMANE, 1902. TEXT S. 29

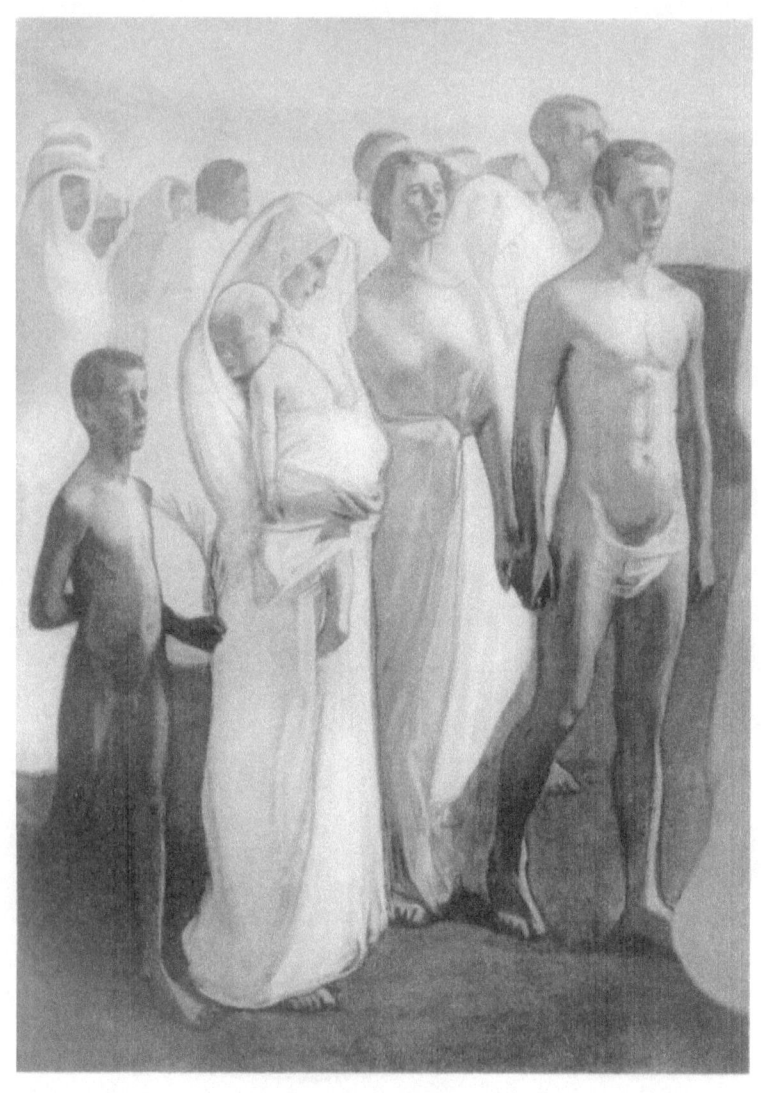

73. MAGNUS ENCKELL, AUFERSTEHUNG. STUDIE ZU DEN FRESKEN IN DER JOHANNESKIRCHE ZU TAMPERE (TAMMERFORS). 1907. TEXT S. 29

74. ANTTI FAVÉN. BILDNIS DES MINISTERS DR. V. WUOLIJOKI. 1924. TEXT S. 29

75. EERO JÄRNEFELT, BILDNIS MATHILDA WREDE. GOUACHE. 1896

76 a. HUGO SIMBERG. DIE TANTE. 1898

76 b. HELENA SCHJERFBECK. SELBSTBILDNIS

77 a. JALMARI RUOKOKOSKI, BILDNIS DES MALERS SALLINEN, 1921. TEXT S. 30

77 b. EERO NELIMARKKA, KNABENKOPF, 1924. TEXT S. 30

78. HUGO SIMBERG, AURORA, 1902. TEXT S. 28

79. ILMARI AALTO, STILLEBEN, 1921. TEXT S. 30

80. TYKO SALLINEN, DER WETTLAUF, 1917. TEXT S. 30

81. TYKO SALLINEN, DIE SEKTE DER HIHLITEN, 1918. TEXT S. 30

82. ALVAR CAWÉN, WIEGENLIED, 1921. TEXT S. 39

63. MARCUS COLLIN, AUF DEM FRIEDHOF, 1914. TEXT S. 30

81. S. W. SIPILÄ, DAS FENSTER, 1931. TEXT S. 30

85. YRJÖ OLLILA. BENZIN. 1928

86. INTO SAXELIN, MÄDCHEN 1928. TEXT S. 32

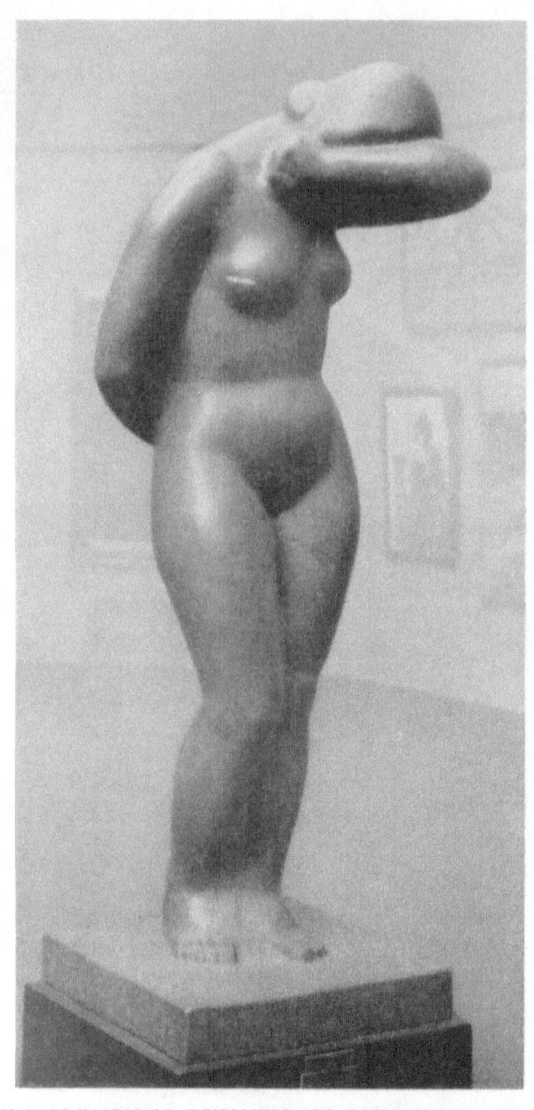

87. JOHANNES HAAPASALO, WEIBLICHER AKT, ROTER GRANIT. TEXT S. 32

88. VÄINÖ AALTONEN, DER LÄUFER PAAVO NURMI. BRONZE. TEXT S. 32

89. VÄINÖ AALTONEN, BRÜCKENFIGUR IN TAMPERE (TAMMERFORS)

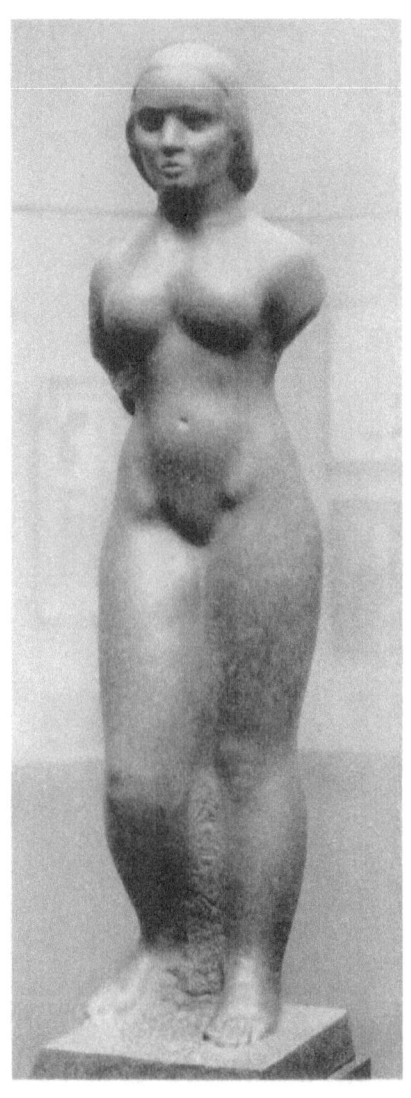

90. VÄINÖ AALTONEN, WEIBLICHE FIGUR. SCHWARZER GRANIT. TEXT S. 32

91. V. JANSSON, WEIBLICHE FIGUR, BRONZE. TEXT S. 32

92 a. FELIX NYLUND. BILDNIS DES MALERS HOLBÖ. BRONZE.
TEXT S. 32.

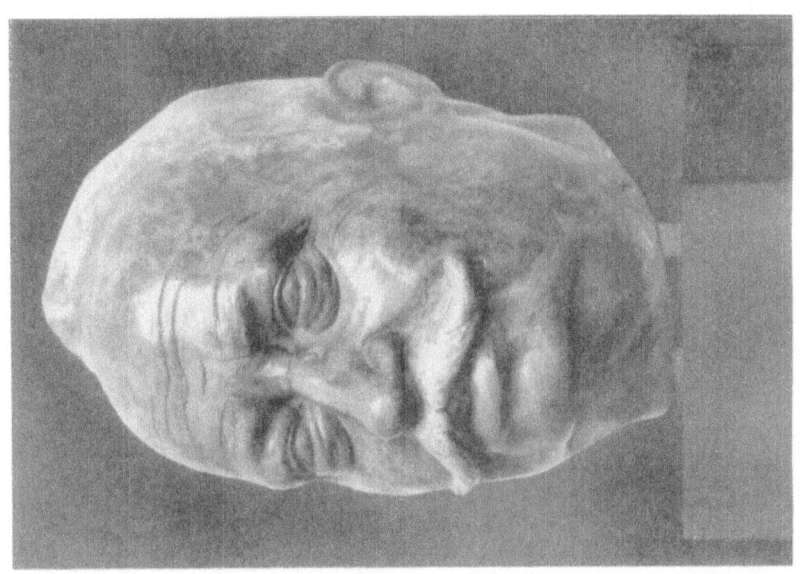

92 b. G. TIGERSTEDT. BILDNIS DES BILDHAUERS VALLGREN.
TERRAKOTTA. TEXT S. 32.

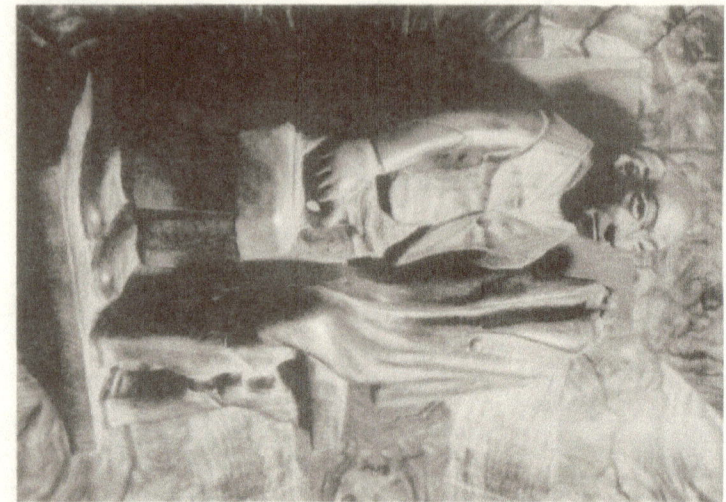

93. HOLZSCHNITZEREIEN VON HANNES ALTERE. TEXT S. 33

94. BAUERNTEPPICH. RYE AUS SAARIJÄRVI 1721. TEXT S. 34

95. BAUERNTEPPICH. RYE AUS PARAINEN 1787

96. EVA BRUMMER: MODERNER GEKNÜPFTER WANDTEPPICH. TEXT S. 34

97. MAIJA KANSANEN, MODERNER GEWEBTER WANDBEHANG. TEXT S. 34

98. BESTICKTE DECKE AUS SATAKUNTA

99 a. TRUHE AUS NORD-POHJANMAA (ÖSTERBOTTEN)
99 b. SCHLAFSOFA AUS NORD-POHJANMAA (ÖSTERBOTTEN)

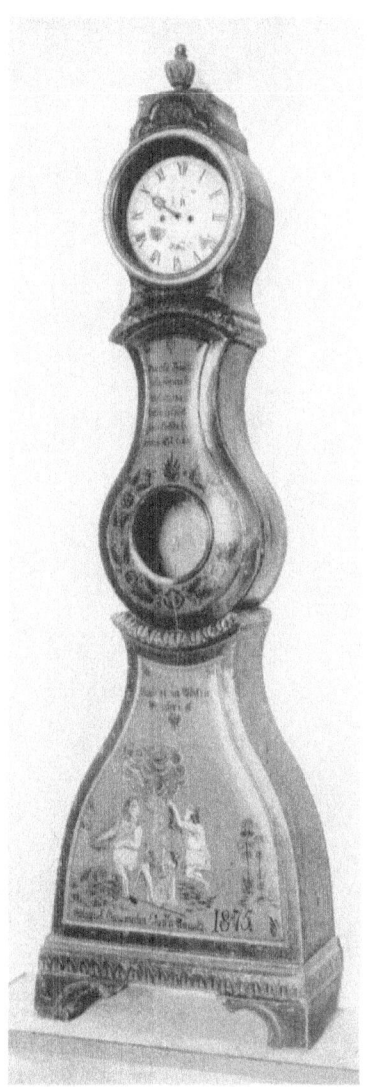

100 a und b. BÄUERLICHE STANDUHREN AUS WESTFINNLAND

101. SCHRANK AUS POHJANMAA (ÖSTERBOTTEN)

102 a. SPIEGEL AUS ÖSTERBOTTEN
102 b. STUHL UND SESSEL AUS ÖSTERBOTTEN UND SATAKUNTA

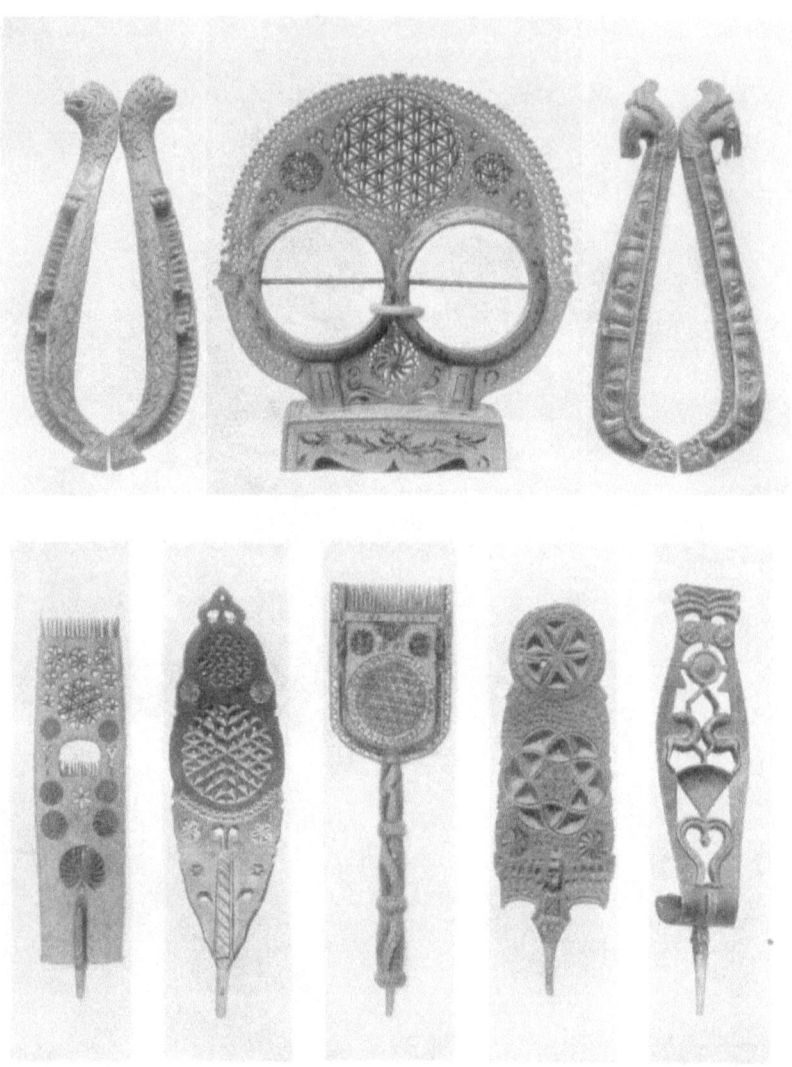

103 a. SPULENHALTER UND PFERDEKUMMETE AUS POHJANMAA (ÖSTERBOTTEN)
103 b. SPINNROCKENAUFSÄTZE. TEXT S. 33

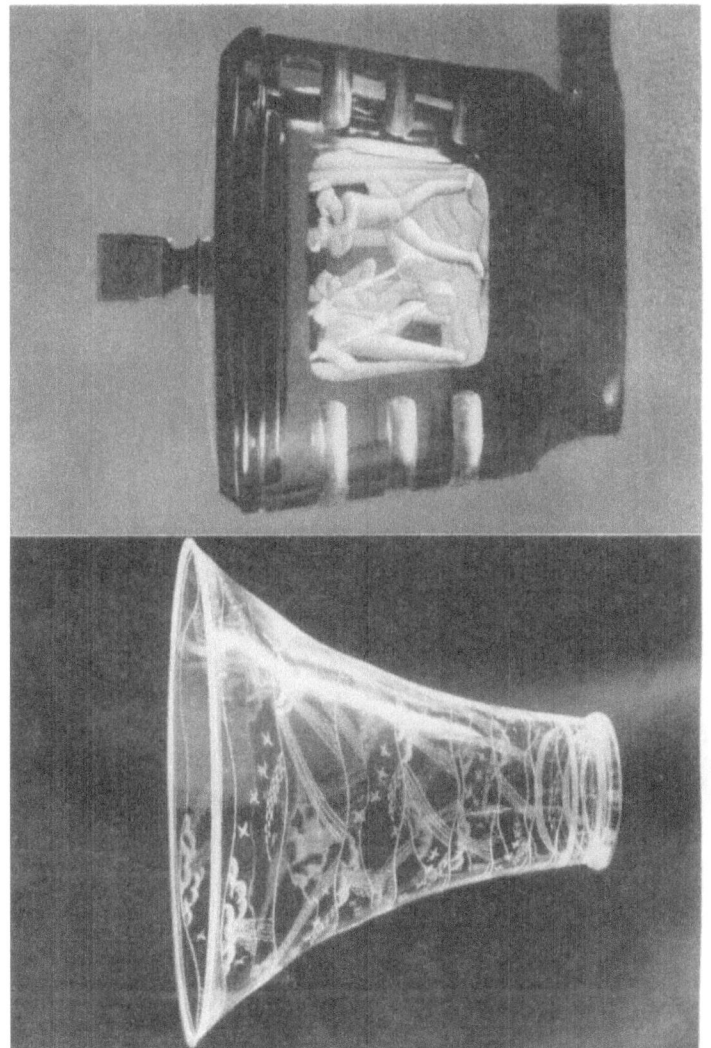

101. MODERNE ZIERGLÄSER AUS DER GLASFABRIK RIIHIMÄKI TEXT S. 35

www.ingramcontent.com/pod-product-compliance
Lightning Source LLC
Chambersburg PA
CBHW031420210526
45464CB00005B/1983

Chelsea Gifford

Colony Collapse Disorder

Chelsea Gifford

Colony Collapse Disorder
The Vanishing Honeybee (Apis mellifera)

Scholar's Press

Impressum / Imprint

Bibliografische Information der Deutschen Nationalbibliothek: Die Deutsche Nationalbibliothek verzeichnet diese Publikation in der Deutschen Nationalbibliografie; detaillierte bibliografische Daten sind im Internet über http://dnb.d-nb.de abrufbar.
Alle in diesem Buch genannten Marken und Produktnamen unterliegen warenzeichen-, marken- oder patentrechtlichem Schutz bzw. sind Warenzeichen oder eingetragene Warenzeichen der jeweiligen Inhaber. Die Wiedergabe von Marken, Produktnamen, Gebrauchsnamen, Handelsnamen, Warenbezeichnungen u.s.w. in diesem Werk berechtigt auch ohne besondere Kennzeichnung nicht zu der Annahme, dass solche Namen im Sinne der Warenzeichen- und Markenschutzgesetzgebung als frei zu betrachten wären und daher von jedermann benutzt werden dürften.

Bibliographic information published by the Deutsche Nationalbibliothek: The Deutsche Nationalbibliothek lists this publication in the Deutsche Nationalbibliografie; detailed bibliographic data are available in the Internet at http://dnb.d-nb.de.
Any brand names and product names mentioned in this book are subject to trademark, brand or patent protection and are trademarks or registered trademarks of their respective holders. The use of brand names, product names, common names, trade names, product descriptions etc. even without a particular marking in this works is in no way to be construed to mean that such names may be regarded as unrestricted in respect of trademark and brand protection legislation and could thus be used by anyone.

Coverbild / Cover image: www.ingimage.com

Verlag / Publisher:
Scholar's Press
ist ein Imprint der / is a trademark of
OmniScriptum GmbH & Co. KG
Heinrich-Böcking-Str. 6-8, 66121 Saarbrücken, Deutschland / Germany
Email: info@scholars-press.com

Herstellung: siehe letzte Seite /
Printed at: see last page
ISBN: 978-3-639-71631-3

Copyright © 2014 OmniScriptum GmbH & Co. KG
Alle Rechte vorbehalten. / All rights reserved. Saarbrücken 2014

Acknowledgements

I would like to acknowledge those who have helped me to find resources for this thesis, who have guided me along the way, and who have inspired the movement for honeybee awareness everywhere. Thank you Dale Miller for encouraging me to do a thesis and for helping me become more engaged about the importance of honeybees through intriguing conversation. Thank you Diana Oliveras and Michael Breed for constructive criticism throughout the process. Thank you Jessie Lucier, Julie Finley, Angie Giustina and my fellow bee guardian group for helping me accumulate resources and for supporting my honeybee awareness events. You have all inspired me and helped me through writing this thesis and have made it enjoyable. Also, a big thank you to Tom Theobald, a beekeeper who has taken his love for bees combined with his newfound knowledge of pesticides to save the creatures that have supported his lifestyle. Keep up the hard work and know you have a growing support system. Lastly, I would like to thank my family for taking the honeybee fight seriously and for integrating it with their own work and conversations. Keep spreading bee awareness!

Table of Contents

SECTION I: A Brief Background on Bees:

Introduction..3
Background..5
    Natural History of Honeybees..6
    History of Hive Evolution and Loss...10
Methods..12

SECTION II: Potential Causes of Colony Collapse Disorder:

Mites and Viruses...13
    Mites..14
        Trachael Mites..14
        Varroa Mites...16
        Mite Treatment...17
    Viruses..18
        Israeli Acute Paralysis Virus..18
    Fungal Diseases...19
        Nosemosis..20
        American Foulbrood..21
        Chalkbrood..22
Human-related Factors...23
    Transportation of Honeybees..24
        Malnutrition...26
    Genetic Loss...27
    Pesticides...28
        Neonicotinoids...29
            Imidacloprid..29
            Clothianidin..30
    Cell Phones..32

SECTION III: Effects of Colony Collapse Disorder:

Economic Effects..33
Ecological Effects...35
Social Effects..35

SECTION IV: Mitigation Strategies for Colony Collapse Disorder:

Policy Recommendations..37
      France..37
      Germany..38
      New York..39
Other Recommendations and Strategies..39
Conclusion..42
Further Research..43

SECTION I: A Brief Background on Bees

"The Creator may be seen in all works of his hands; but in few more directly than in the wise economy of the Honey-Bee."
~L.L. Langstroth~

Introduction

Honeybees have been disappearing in large numbers across the globe. Where are all of the honeybees going, and why? Honeybees are considered a keystone species because of the significant role they play in supporting various ecosystems through their vast pollination services. Humans are also extremely dependent on these pollination services, which begs the question: How severely will human civilization be impacted if honeybee populations continue to decline? Not only do honeybees produce wax and honey, they are also directly and indirectly responsible for pollinating a third of the food that humans consume. Their ecological and economic contributions are invaluable, which makes colony collapse disorder a very important topic of discussion. Although government researchers, scientists, and beekeepers have been trying to find the answer to why honeybees are disappearing, no single factor has been determined. However, if the disorder remains untreated, continued honeybee loss could drastically affect America's food supply as well as the larger agricultural, economic and environmental systems.

Colony collapse disorder (CCD) is a phenomenon in which worker bees abruptly disappear from a beehive. Since CCD was first reported on the East coast of the United States in 2006, continued increases in honeybee loss are making CCD an extremely pressing issue. Symptoms of colony collapse disorder include the rapid loss of adult worker bees, few or no dead bees found in the hive, only a small cluster of bees with a live queen present, and pollen and honey stores remaining in the hive (Debnam 2009). So far, there has not been a conclusive answer as to why seemingly healthy bees are disappearing. Scientists are using a variety of methods to evaluate this collapse disorder to try to find what is responsible for the malady.

There are many theories surrounding the recent honeybee population decline. Natural stressors include various parasites and pathogens, such as fungi, viruses, and mites, all of which can be extremely harmful to bees. Some of the anthropogenic factors that are most likely contributing to collapse include: increased exposure to pesticides, trucking honeybees across long distances for the pollination of commercial crops, poor nutrition, artificial insemination of queens using sperm of limited genetic variability, and habitat loss. It has even been suggested that increased cell phone use may be interfering with the honeybees' immune system and/or their ability to navigate. This thesis will bring together knowledge about the possible reasons behind the loss of honeybees. It will look into the science and the controversies surrounding this crisis.

In addition to exploring the possible causes of CCD, this research will also highlight the effects of this disorder. As a keystone species, honeybees play an important role in the ecosystem at large. These insects are anatomically constructed to work symbiotically with flowering plants to receive pollen and, in turn, facilitate pollination. The mutualistic actions by these two groups of organisms maintain both plant and animal diversity in the environment. Additionally, honeybees are considered an index species because the health of the species indicates the condition of the environment. Honeybees are extremely susceptible to change, and as more pathogens and poisons are introduced into the environment, the more adversely affected these pollinators will be. CCD has significant impacts on the environment and will lead to dramatic repercussions for natural and human systems that depend on the products and services that honeybees provide.

The impacts of colony collapse are an urgent matter. On a small scale, the livelihood of beekeepers depends on the viable supply of honeybees and their role as pollinators and honey producers. With as much as 30-90% losses in a single winter, beekeepers do not have a viable population to sustain themselves economically for future years (Kaplan 2008). On a larger scale, major declines of bee populations will force the Untied States to rely on food imports from other countries that have not been subject to such great losses. This means that all fruits, vegetables, meat products, coffee, teas, and other staples on which we rely would have to come from other places. The increased price of food and elevated trade deficit, coupled with significant job losses in the agricultural sector, would further cripple the American economy. Eventually, if strategies are not implemented to better understand and mitigate the causes and effects of colony collapse, we could see one of the biggest natural and economic disasters of our time.

The importance of this issue should not be underestimated. Education and research about the larger role that bees play in nature and our human connection to that system need to be strengthened. It is also critical for governments and communities to implement various policies and programs that will protect the livelihood of honeybees as a species and ecological partner. With around 130 agricultural plants in the United States dependent on honeybee pollination (valued at $15 billion annually in the United States and $215 billion worldwide), the disappearance of bees is a crisis we cannot afford to ignore (McGregor 1976, VanEngelsdorp et al. 2008).

Background

Honeybees (*Apis mellifera*) possess traits that make them successful as both pollinators and cooperative social creatures. Their bodies are constructed to work in perfect harmony with pollen producers. Honeybees benefit the larger ecosystem by facilitating pollination, thereby allowing plants to grow, bear fruits and seeds, and provide nutrients for other animals. Human beings have also relied on honeybees for their capacity to produce wax and honey. Understanding the traits that honeybees posses is important to appreciating the contributions that these insects provide to both humans and the ecosystem as a whole.

Natural History of Honeybees

Ancient Egyptians worshipped honeybees and believed that bees grew from the tears of Ra (Readicker 2009). As early as 3500 BC, a hieroglyph of a bee represented the King of Lower Egypt, a symbol that lasted more than four millennia. Honey was considered a tribute and wax was used to make sacred figurines. In India, Vishnu described the bee as "the creator and destroyer of all existences, one who supports, sustains and governs the Universe and originates and develops all elements within" (Readicker 2009). In Greek mythology, honey was the food of the Gods (Schacker 2008). Regardless of the specific civilization, honeybees have been considered a treasure, a gift, a blessing, and a creature to be respected.

Bees have been in existence for roughly 100 million years, since the earliest flowers developed; the honeybee evolved as a subspecies roughly 35 million years ago (Readicker 2009). Although there is not a complete record of the history of honeybees, the relationship between bees and flowering plants is undoubtedly one of symbiosis. Today, there are 250,000 flowering plants, and 20,000 of them rely on bees for pollination (Schacker 2008).

Honeybees are built perfectly to be successful pollinators. They have two different sets of eyes, the ocelli and the compound eyes. The ocelli consist of three eyes, each with a dense lens. They are used to detect the intensity of light, which helps them to remain right side up (Winston 1987). The other set consists of two compound eyes, each with 6,900 hexagonal lenses that are able to decipher different light conditions, colors, and sun position. Hairs on their body are used to detect wind speed and direction, and their sensitivity to ultraviolet light directs them to the plant's nectar and pollen storage (Benjamin 2009).

Honeybees have antennae on the top of their head that act as smell detectors (see Figure 1; Snodgrass 1984). Bees are 100 times more sensitive than humans to odors such as flowers, nectar wax, and propolis (a plant resin used to seal honeycombs). Honeybees also have a folding tongue called a proboscis that allows them to reach deep into flowers for nectar. Their mandibles hold and manipulate wax and collect propolis (Snodgrass 1984). Other functions of the mandibles include ingesting pollen for food, cutting, cleaning, grooming, and fighting (Winston 1987). Their feet have hooks in order to hang onto flower petals and pads that allow them to walk upside down on flat surfaces. Their front legs are also equipped with hooks that are used to clean the antennae, and their middle legs act as wax collectors (Benjamin 2009). Once the wax is collected, it is passed to the front legs for the mandibles to manipulate into comb. When a honeybee lands on a flower, the pollen is brushed into the corbicula, which is located on the hind leg and is capable of holding eight milligrams of pollen each time the bee flies (Benjamin 2009).

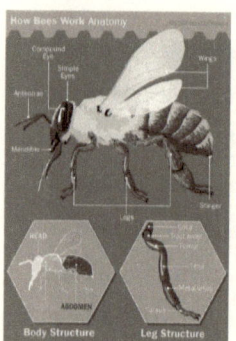

Figure 1

Honeybees make thousands of visits to one type of flower at a time until the food source has dried up. In a single day, one worker bee will visit, on average, 1,500 flowers to gather one load of pollen (Benjamin 2009). In order to produce one gallon of honey, a hive will collect pollen and nectar from around 500 million flowers, and will have flown around seven million miles (Schacker 2008). According to the USDA Agricultural Statistics Service, the average honey yield per colony was around 65 pounds in 2006, although some colonies can produce around 200 pounds (USDA 2007, Hinke 2007). The amount of honey produced is dependent upon the health of the colony and the availability of forage and flowering plants.

Foraging is crucial for pollination, the collection of food for colony, and the production of wax and honey. Honeybees are directionally sensitized and use the sun and landmarks as points of reference and can travel three miles or sometimes further from the hive in search of food without losing track of where the colony is located (Benjamin 2009). When a honeybee returns to the colony after discovering a rich food source, it will do a "waggle dance" in order to relay directional cues to the rest of the worker bees in the colony. This dance entails shaking the abdomen side to side in order to communicate to the other bees the flying time to a food source, the wind speed, and the sun's direction. If the dance is merely a fraction of a second, the source is nearby, but if it lasts a couple of seconds, the source is about five minutes away (Tautz 1996). Encoded in the dance are scent clues indicating what the food is and how much of it there is (Readicker 2009). The bee will repeat this dance a number of times in order to convey its message to the other workers. If there are not enough workers that leave the colony to forage, the bee will conduct a "tremble dance" to recruit more workers (Benjamin 2009).

As the honeybees visit flowers, they facilitate pollination. They will visit many of the same species of flowers in a single trip and their hairy bodies pick up pollen grains at each stop. Bees pollinate 16% of the world's flowering plant species and 400 of the world's agricultural plants (Delaplane 2000). Bee foraging is advantageous for the flowering plants because they encourage high rates of pollination and plants, in turn, flowers provide nectar and pollen for their honeybee pollinators to produce wax and honey (Delaplane 2000).

Honeybees are social insects, meaning that a division of labor within the hive is a fundamental aspect of efficient organization. In a colony, a bee can be a queen, a male drone, or a female worker. Each bee has its own role within the hive and is responsible for reproducing, bringing in food, or cleaning (Kolmes 1989). Honeybees go through three stages in the process of becoming an adult: egg, larva, and pupa. Together, each of these stages is referred to as the brood. Unfertilized eggs turn into male drones and fertilized eggs mature into either female workers or queens (Readicker 2009).

There is only one queen in a colony at a time, and her primary role is reproduction. She is equipped with ovaries, queen pheromones and a barbless stinger, and she is substantially larger than the rest of the bees (MAAREC 2011). It is her job to lay eggs in order to populate the colony. If she is healthy and has ingested enough food, she will lay about two thousand eggs per day and live for five to seven years (Schacker 2008). Another important duty of the queen is to produce pheromones to keep the colony together. Honeybees identify themselves by the smell of the colony, which the queen secretes through her mandibular glands (MAAREC 2011). If a larva is destined to be a queen, she is consistently fed royal jelly by the worker bees, which is a white, protein-rich substance that turns her into a "sexually mature powerhouse" (Meyerowitz 2000). As a queen loses her fertility, a new chamber is created for a successor. During the summer, a number of queens are raised in an individual colony, and the ones that do not take over the hive fly off in search of a new home (Benjamin 2009).

If a bee is destined to be a worker, it will take her 21 days to emerge from a pupa (Schacker 2008). Worker bees make up 99% of the colony and have many jobs, the primary one being to forage for nectar, pollen, water, and propolis. Individual worker bees are equipped with brood food glands, scent glands, wax glands, and a pollen basket (MAAREC 2011). Members that are not fully grown have alternate tasks such as cleaning and ventilating the hive, taking care of the brood, receiving nectar and pollen to make royal jelly, building combs, and guarding the entrance (Benjamin 2009). These are all necessary tasks that keep the colony organized and free of diseases.

If a bee is destined to be a drone, it will take him 24 days to emerge (Benjamin 2009). These male honeybees are the biggest in the colony, but they do not have a stinger, pollen baskets, or wax glands. Their job is to fertilize the queen on her mating flight (MAAREC 2011). It takes three to seven days in order for the queen to become impregnated and once she mates, she is equipped with enough sperm to continually populate the colony over the course of her lifetime. The drone dies instantly after he completes his mating duties (Schacker 2008).

The specialization within honeybee colonies demonstrates the cooperation within a functioning and efficient system. However, as various pathogens infiltrate the honeybee's system, the integrity of the colony can weaken. This is why it is crucial to have a healthy queen and a healthy colony to get through times of infection. If colonies get smaller, the chance of disease increases, making the colonies more susceptible to collapse.

History of Hive Evolution and Loss

Honeybees are not native to the United States, but were first introduced to North America when Europeans crossed the Atlantic and facilitated the species' migration in 1621 (Horn 2008). The natural habitat of honeybees covers a vast area. It ranges from the tip of Africa and the Mediterranean and extends to northern Europe and southern Scandinavia (Winston 1987). Subspecies of honeybees have been determined by the different adaptations they have acquired throughout a variety of climatic conditions. Honey has been used as food ever since the first *Homo sapiens* lived in caves (Benjamin 2009). The first beekeepers can be traced back to 2,400 BC when Egyptians transported their bees to produce honey (Benjamin 2009). Beekeeping spread westward along the North African coast and reached parts of Greece and Rome where beekeepers further disseminated the information known about honeybees (Crane 2004). Throughout Zimbabwe and South Africa, images can be found of hunters obtaining honey by using smokers to lure the bees out of their nests. Honey hunting dates back 15,000-20,000 years, indicating how useful honeybees have been to humans throughout history (Benjamin 2009).

Honeybees were an important element of the colonization of America. As European settlers came over and planted seeds and saplings, the bees were necessary for pollination. Bee pollination resulted in the increased spread of clover, which was

Figure 2

used to feed the livestock (Benjamin 2009). The settlers continued to spread honeybees by aiding their move across plains and mountainous regions. The production of honey sharply increased during the 17th century and by 1730, Virginia was exporting around 344,000 pounds of beeswax (Benjamin 2009). Beekeeping became even more commercially feasible during the 19th century as certain tools were invented and made accessible. This equipment included the smoker, the frame hive (Langstroth hive, see Figure 2), the honey extractor, and comb foundations (Horn 2008). By the 20th century, the western honeybee became a vital part of human existence.

However, as humans continued to manipulate the honeybee and deliberately transferred them on a global scale, diseases simultaneously spread and harmed managed colonies. Colony losses have occurred periodically throughout history. Fungus, mites, and starvation have all been thought to be the cause of the deaths. The first recorded collapses were called "May Disease" in Colorado in 1891 and 1896 (Schacker 2008). Investigations pointed to a fungus, *Aspergillus flavus,* as the culprit of the Colorado collapses. *Aspergillus flavus* causes stonebrood, a fungal disease, which targets both the immature and adult bees. Since adults are affected, it is difficult for healthy bees to carry infected bees out to reduce spreading of fungus (Underwood 2010). Another epidemic occurred on the Island of Wight in the United Kingdom. Between 1905 and 1919, 90% of the bees died and the reason as to why is still debated (Neumann 2010). In the twentieth century, large-scale losses occurred throughout North America (Benjamin 2009). Numerous studies have been conducted to determine the causes of these losses but no definite reason has been found.

In 2006, Dave Hackenberg, a commercial beekeeper residing in Rushkin, Florida, reported what we now call CCD. He was first referred to as an unskilled beekeeper, but once a third of all the honeybees in the United States disappeared, people thought otherwise (Benjamin 2009). Hackenberg lost 2,000 of his 2,950 hives and rebuilding was extremely costly. As seen in Figure 3, by 2007 more than 22 states had reported CCD with some beekeepers losing up to 95% of their hives. Losses continued to grow, as 35 states reported CCD by 2008 (Schacker 2008). These dwindling numbers of bees are now being seen as a crisis.

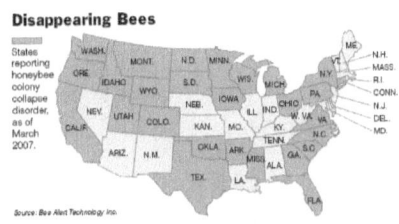

Figure 3

Honeybee pollination is vital for commercial agriculture. Without them, the agricultural sector will be significantly affected. Approximately a third of human food is directly or indirectly supported by honeybee pollination (Kremen et al. 2009). Due to the immense importance of honeybees, there needs to be further research on this pressing issue. This thesis will discuss the many factors that may be causing colony collapse as well as potential mitigation strategies and areas of further research.

Methods

Scholarly primary and secondary sources were used to gather information on CCD and honeybees. Sources included journal articles and books outlining information on honeybees and colony collapse disorder. By researching multiple factors individually as well as compilations of information, the timeline of events connected to the disappearance of hives across the world has become increasingly clear.

Knowledge of the repercussions of colony collapse disorder was also gained by attending beekeeping courses, meeting with beekeepers in Boulder County (Colorado, USA), going to events to educate the public on CCD, and participating in open forums at the Boulder municipal building focused on pesticide use within the city. As an addition to my research, I contributed the knowledge gained through writing this thesis to promote public education on colony collapse. One way I did this was by organizing an event to show the film *Vanishing of the Bees* at the University of Colorado at Boulder in the effort to engage students on this pressing issue. Another way I educated my college campus about the effects of pesticides on honeybees was to insert a description of CCD into *The Turf Management Task Force Summary*, a plan to eliminate the use of synthetic pesticides on CU's campus grounds. Spreading information about CCD is the first step to reversing the effects of further honeybee loss.

By accumulating knowledge both in person and through library research, I better understand how colony collapse disorder is being portrayed to the general public, the beekeepers, and the government organizations that will have to make important decisions on how to handle this potential disaster. I hope to increase interest and a better appreciation for honeybees among the general population. Once people understand the fundamental role that these pollinators play, I trust people will want to change their habits to accommodate the backbone of pollination and to encourage their elected representatives to do the same.

SECTION II: Potential Causes of Colony Collapse Disorder

"The apple trees bloomed but no bees droned among the blossoms, so there was no pollination and there would be no fruit."
~Rachel Carson~

Mites and Viruses

Diseases are problematic for nearly all organisms, but they are particularly a problem for social insects. Honeybee colonies provide a favorable environment for parasites and pathogens because of the warm temperature and the high concentration of hosts (Tarpy 2006). Past colony losses have been associated with mites and viruses, therefore making them important to discuss as stressors linked to CCD. Mites have infested honeybee hives long before severe honeybee loss began. However, mite infections may be having an even deadlier impact on honeybees now that there are a variety of other factors that have been introduced. Increased pesticides loads, transportation stress, and genetic loss have become more prominent and these stressors may be combining with mite and viral infections to weaken the bees' immune systems.

Mites

Varroa mites (*Varroa destructor*) and tracheal mites (*Acarapis woodi*) are parasites known to damage colonies (Genersch 2010). Both of these species of mites were introduced to the United States in the 1980's and are now widespread in managed colonies (VanEngelsdorp et al. 2009). Prior to their introduction, beekeepers reported 5-10% winter losses, but as mites spread, these losses rose to 15-25%. This trend of loss due to mite infestation has continued to increase (VanEngelsdorp et al. 2008). Although it seems as if mites are a primary cause of colony collapse disorder, according to a recent study that examined colonies with and without symptoms related to CCD, not all colonies showing signs of significant loss were attributed solely to mite infestation (VanEngelsdorp et al. 2009). Healthy hives and unhealthy hives have mites. This suggests that these parasites may be combining with other factors to synergistically weaken colonies.

Tracheal Mites

Tracheal mites, *Acarpis woodi,* live in the tracheal tubes of honeybees. They prevent infected bees from working as hard or from living as long as healthy bees (Caron 2000). The mites pass quickly from bee to bee and eventually can pass between colonies. Tracheal mites were first discovered in the early twentieth century in Europe where the disease spread rapidly (Sammataro 2000). Since then, the infestations have become increasingly common.

Mites have a fast moving life cycle, with complete development taking place in 11-15 days. Mites are nourished by bee hemolymph, which they obtain by sucking it out of a bee's tracheae (Donzé 1994). Younger bees are more susceptible to these mites because they have a better chance of living long enough for the mite to complete its life cycle. Once a female mite is embedded within the host, she lays eggs in the tracheae (Sammataro 2000). As the bee gets older, its tracheal mite load increases and may force the insect to innately leave the hive in order to prevent the spreading of the infection (VanEngelsdorp et al. 2009).

Controlling the spread of tracheal mites is difficult when there are vast numbers of them in a hive. Mites can lead to smaller bee populations, cause increases in honey consumption, and, in turn, lower honey yields (Sammataro 2000). As the bee populations decrease, the possibility for the winter bees to make it through the cold season diminishes as well. Winter is also when the tracheal mites are the most prevalent because the bees are obligated to stay inside the hive (Sammataro 2000). Studies show that once a colony has tracheal mites, it will remain infested (Caron 2000). The tracheal mite is difficult to detect, so the common thought process among beekeepers it to assume the colony has mites and to treat it aggressively. If they do not, the queen may become susceptible as she ages and the whole colony will collapse (Caron 2000).

Symptoms of tracheal mite infection include declining populations, bees of a weakened state crawling on the ground with disjointed hind-wings, and abandoned hives full of honey (Readicker 2009). The best way to diagnose a honeybee for tracheal mites is to dissect it and look inside the tracheae. Once a hive is infested with mites, there are different ways to treat the parasites. The most common way to treat mites is to spray insecticides. Since both of these insects are similar physiologically, the difficulty is finding a toxin that will only harm the mite, not the bees (Sammataro 2000). Mites have caused the amount of pesticides used to rise in order to rid the colony of infestation. Beekeepers have been experimenting with how to rid the hives of the mites without using chemicals at all. More natural ways to prevent and treat mites need to be encouraged. These more organic methods will lower the amount of harmful pesticides released into the environment and give bees an opportunity to gain a resistance to pests through natural selection. Increasing poison loads is only a temporary solution.

Varroa Mites

The varroa mite, *Varroa destructor,* is the primary pest of the honeybee and is now found in all parts of the world except for Australia, New Zealand, and Hawaii (Sammataro 2000). It was first discovered in Java and while limited to Southeast Asia, it was a pest to primarily the Asian honeybee, *Apis cerana* (Readicker 2009). Throughout the twentieth century, the mites spread through commercial transportation and began affecting colonies of all bee types worldwide. By 1994, 98% of the wild honeybee population was destroyed by mites and the various diseases that they bring with them (Schacker 2008).

Varroa mites are major pests to the honeybees. They can hinder the ability of the queen to reproduce, which can then be fatal to the colony at large (Johnson 2007). They can also attack both female worker bees and male larvae, although they are built to better parasitize the drone cell (Donzé 1994). Mites attach to adult bees by piercing the drone or the worker bee's abdomen or behind the head and ingesting the hemolymph. Mites can enter into the cells and hide in the liquid brood food before the cell is capped. Once the cell is capped, the mite begins laying her eggs on larvae (Sammataro 2000). Depending on the number of mites, the bee can be severely weakened, perhaps fatally. If no preventative measures are taken against the mites, a whole colony can be killed due to either direct effects of the mites or the viruses that they spread (Tarpy 2007).

Some symptoms of varroasis include visible dark mites within the white pupae (see Figure 4), punctured holes on the drone or worker brood cells, or deformed adults crawling around inside or outside of the hive (Sammataro 2000). It is difficult for a beekeeper to determine why a colony may appear weak because various factors may produce similar symptoms. For example, the effects of toxins within a bee's system may appear the same as a mite-infested colony, making colony collapse difficult to examine.

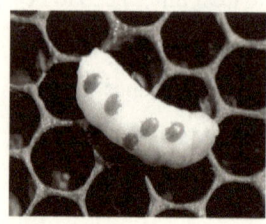

Figure 4

Mite Treatment

To treat mites, many beekeepers use miticides, such as fluvalinate (Apistan®), a pyrethroid, or coumaphos (Bayer CheckMite ™), which are applied to the hive as plastic strips (Sammataro 2000). Mites, however, are beginning to develop a natural resistance to the chemicals, thus making it more difficult for beekeepers to control them (Ambrose 2000). The use of essential oils, smoke, and traps are alternate, more organic methods of controlling mites, and many beekeepers are increasingly choosing these over synthetic insecticides. Their effectiveness, however, may depend on the number of mites within the hive (Ambrose 2000). Although mites and the chemicals to treat them may not be the direct cause of CCD, the increased dosages of toxins may result in worse problems for bees.

According to Pat Flinn, a beekeeper in Alberta, Canada, varroa mites have become a worse problem than ever before (Flinn 2011). Some beekeepers in the area have lost as many as 50% of their bees in the winter months. The provincial government has sent inspectors to look at hives in order to identify the problematic mites and accompanying diseases. In addition, workshops on Integrated Pest Management (IPM) have been established to educate beekeepers on the procedures that should be used to monitor their hives and treat for mites. Flinn uses Apivar®, (Amitraz, formic acid) strips like many beekeepers worldwide whose hives have suffered from mites. Other insecticides include formic acid and Mite Wipe absorbent pads, which can be used to treat tracheal mites as well as varroa mites. CheckMite and Apistan strips can also be obtained, she said, but there are strong concerns about resistance and residues left in the wax. According to Pat, there is too much fear of losing hives by not using these chemicals.

Although treating symptoms with pesticides seems to be a common solution, other solutions need to be researched so that the mites do not develop resistance. When insecticide resistance happens, mites essentially turn into "super" mites and stronger toxins will need to be used. Such escalation in drug use may compromise the health and immune system of the bees (Schacker 2008). It is critical to continue research on mites and the procedures we use to treat them, because the synergistic effects may be fatal to honeybees.

Viruses

It is important to study viruses in relation to mites because mites are often carriers of viral pathogens and are able to directly inject them into the bee's hemolymph (Tentcheva et al. 2004). The only virus that will be discussed in this thesis is the Israeli Acute Paralysis Virus because it has been commonly discussed as a predictive factor of CCD.

Israeli Acute Paralysis Virus (IAPV)

Israeli Acute Paralysis Virus (IAPV) was first described in 2004 in Israel and was said to be imported from Australia in 2005 (USDA 2008, Berry 2007). IAPV can affect both brood and adult bees and is transmitted by varroa mites (Sammataro 2000). The virus is spread when mites feed on honeybees. Other ways the virus is transmitted is when honeybees have physical or mouth-to-mouth contact with an infected bee or if a nurse bee cleans up virus-infested feces (Sammataro 2000). Symptoms of Israeli Acute Paralysis Virus include shivering wings and eventually paralysis and death outside of the hive (Schacker 2008).

IAPV research has proven that the virus is not a recent introduction from Australia and has been present in the United States since 2002 (USDA 2008). The virus can appear in hives with or without CCD, meaning that the correlation between the two may not be significant (Schacker 2008). However, phylogenetic analyses have shown that IAPV is an important pathogen that should be considered as a candidate for what is causing CCD (Palacios et al. 2008).

The type of colony management may also be another factor to consider. Large commercial beekeepers have been more affected by IAPV than small-scale beekeeping operations, suggesting that the stress of the honeybee, pesticides, or poor nutrition may be correlated with weakened immune systems and infection (Readicker 2009). Further tests are being done to determine if viruses should still be considered a relevant factor in explaining CCD and the differences between imported and domestic strains of IAPV (Kaplan 2008).

According to Eric Mussen, an entomologist from UC Davis, "Honey bee colonies are being stressed by a large number of things and when those stresses become overwhelming, the bees simply fly away from the hive" (Mussen 2011). Researchers who study colony collapse disorder are exploring viruses, such as IAPV, in combination with other stressors in order to understand why honeybees are becoming more susceptible to disease.

Fungal Diseases

Some of the honeybees that have been recovered and dissected have shown high levels of bacteria and fungi. Nosema and other pathogens such as American foulbrood and chalkbrood can contaminate hives. Some researchers have suggested that once honeybees are infected with disease, their immune systems are compromised and they are more likely to disappear from the colony (Johnson 2009). On the other hand, immune deficiencies caused by other stress-induced factors may be causing honeybees to become more vulnerable to infection.

Nosemosis

Nosemosis is caused by a single-celled fungus (microsporidia) that is an obligate intra-cellular parasite (Paxton 2010). The parasite's only means of multiplying is by living in honeybee mid-gut cells, where it takes over the cell's functions of obtaining nutrients. It eventually ruptures and kills the cell and is then released into the gut lumen and into other mid-gut cells (Schacker 2008). A *Nosema apis* infection prevents the development of the glands that secrete royal jelly, leading to sick bees unable to feed the brood (Vivian 1986). Queens who get infected usually stop producing eggs, slowing down growth of the entire colony (Somerville 2007). The parasite is spread easily through defecation by infected bees because other healthy bees, which naturally clean the hive, ingest the parasite. The hive will have dark and yellow streaks on the front and on the comb (see Figure 5; Vivian 1986). The bees will twitch and will be unable to fly due to disjointed wings. They will also have enlarged abdomens and lack body hair (Vivian 1986).

Figure 5

Nosema ceranae was reported in Taiwan in the spring of 1995, when the virus had transferred from the Eastern honeybee (*Apis cerana*) to the Western honeybee (*Apis mellifera*) (Paxton 2010). The parasite spread and was eventually discovered in Europe in 2006. Although the bees suffered from nosemosis, nosema is still not

20

described as causing symptoms that are commonly associated with CCD (Schacker 2008).

While nosemosis is not considered to be the cause of CCD, it is still viewed as a major stressing factor that can infect hives and potentially kill the colony (Paxton 2010). A major difference between *N. ceranae* and *N. apis* is in the severity of the infection. *N. ceranae* is more aggressive and can occur year round, collapsing a hive in merely eight days (Schacker 2008). *N. apis,* on the other hand, usually only appears in the winter months when the bees are under more stress. It often goes away once the spring comes and foraging begins again (Somerville 2007).

One way to treat nosemosis is by using an anti-parasitic drug called Fumagilin-B®. The chemical is effective against the parasite within the gut of the bees, but not against the spores (Vivian 1986). A strong colony can resist nosemosis. However, when honeybees are under increased stress, they are more prone to infection.

American Foulbrood

American foulbrood (AFB) is a crippling disease caused by *Paenibacillus larvae ssp.,* a spore-forming bacterium (Spivak 2001). Young larvae no more than three days old are the most susceptible to this particular disease and can be infected by feeding on spores (Antúnez et al. 2009). The spores then develop inside of the larvae's guts (Vivian 1986). Once a honeybee is infected, both the population and the overall production (e.g. honey, pollen, propolis, royal jelly, and beeswax) decrease (Spivak 2001). When the bacteria infect a larva, the larva first turns a tan color. Death makes the larva look dark brown and shriveled up (Antúnez et al. 2009). An irregular brood pattern may be another sign that a hive is infected (Conrad 2009). This irregularity may be evidenced by a number of empty cells among capped cells meaning that the brood is not being formed at the same time. Lastly, there will be a foul smell emanating from the infected combs (Conrad 2009).

AFB is one of the most serious and common diseases for honeybees. Its spores can remain viable for up to fifty years waiting for the perfect environment for growth and reproduction (Antúnez et al. 2009). Once the spores germinate, they contaminate honey and comb as well as permeate into wood fibers. AFB is easily transferred between bees within a colony as well as between colonies (Vivian 1986). This is why it is imperative that the beekeeper is immediately aware of the signs indicating AFB.

American foulbrood is a worldwide outbreak and two percent of all colonies have it at a given time (Vivian 1986). Treatment includes burning the bees, combs, and the equipment, or using antibiotics. Burning seems to be an effective method, but it is an extremely destructive approach. Antibiotics only work against active bacteria, not spores (Antúnez et al. 2009). The spores still remain in the honey, wax, and pollen and may infect the hive again when the antibiotic is removed (Conrad 2009). American foulbrood does not lead to colony collapse by itself, but additional studies need to examine if the chemicals that are used to combat AFB are further weakening the bee's immune system making them more susceptible to infection (Conrad 2009).

Chalkbrood

Chalkbrood is a fungal disease caused by *Ascosphaera apis* (Vivian 1986). It was first reported in California in 1968 (Vivian 1986). Fungal spores are ingested by the bee and the growing fungal mycelium causes the infected larva to mummify and turn white. If the mycelia are of opposite sexes, the mummies will turn a darker gray color (Gilliam 1983). Larvae that are about to be sealed or have just been sealed are the most vulnerable to the disease (Flores 2005). Pollen combs may be a primary source of transmittance and if infected pollen combs are transferred to healthy hives, the healthy brood will ingest the spores (Flores 2005). Further research needs to be conducted to better understand how to prevent the conditions that make brood susceptible to chalkbrood.

Researchers have found *Nosema ceranae,* chalkbrood, and other pathogens in infected hives suggesting that high levels of contamination are severely harming the immune system of honeybees (Johnson 2009). However, natural resistance to these particular pathogens is more successful when a honeybee colony is in good condition. When colonies receive an adequate food supply and the bees within that colony are healthy and maintain hygienic grooming, the colony will have a higher resistance to disease. However, as honeybees are weakened by a variety of other destructive anthropogenic factors, less hygienic grooming takes place, meaning the colony will have less of a resistance to disease.

Increased stress is most likely leading to the honeybee's weakened immune system and the bee's frail state may be allowing infection to take over the hive. Scientists, researchers, and beekeepers are still trying to correlate the symptoms of natural causes such as mites and pathogens with the stress induced by anthropogenic causes. If all of the potential causes of CCD are well understood, then we can begin to find ways to reverse the dramatic honeybee losses.

Human-related Factors

Anthropogenic causes of CCD include: transportation of honeybees for commercial pollination, malnutrition, genetic loss due to artificial queen insemination, and increased amounts of pesticides used on colonies and in the environment. Although mites, viral infections, and pathogens can cause honeybee fatality and diminish the population within a colony, disease does not explain the unprecedented widespread losses that are occurring today. There must be other factors that are contributing to the colony's weakened state. It is mandatory to understand all of these potential stressors in order to better grasp the complexity of colony collapse disorder. This knowledge will be what drives a change in human behavior that is vital to minimizing the future loss of pollinators.

Transportation of Honeybees

Western honeybees are considered one of the most valuable agricultural pollinators because they can be transported easily with relatively little maintenance (Williams et al. 2010). Industrial-scale beekeeping requires hundreds of thousands of hives to be trucked around the country for short seasons of pollination (Stokstad 2007). According to the Food and Agriculture Organization (FAO), the global population of commercial honeybee colonies has increased by 45% since 1961 (Aizen 2009). This statistic suggests that fast-paced economic globalization has been raising the demand for managed hives as a service for agricultural pollination. However, this increased demand is occurring more rapidly than the supply of managed hives, which may be stressing the global pollination capacity (Aizen 2009).

Table 1. Estimated Value of the Honey Bee to U.S. Crop Production, by Major Crop Category, 2000 Estimates

Crop Category (ranked by share of honey bee pollinator value)	Dependence on Insect Pollination	Proportion of Pollinators That Are Honey Bees	Value Attributed to Honey Bees[a] ($ millions)	Major Producing States[b]
Alfalfa, hay & seed	100%	60%	4,654.2	CA, SD, ID, WI
Apples	100%	90%	1,352.3	WA, NY, MI, PA
Almonds	100%	100%	959.2	CA
Citrus	20% - 80%	10% - 90%	834.1	CA, FL, AZ, TX
Cotton (lint & seed)	20%	80%	857.7	TX, AR, GA, MS
Soybeans	10%	50%	824.5	IA, IL, MN, IN
Onions	100%	90%	661.7	TX, GA, CA, AZ
Broccoli	100%	90%	435.4	CA
Carrots	100%	90%	420.7	CA, TX
Sunflower	100%	90%	409.9	ND, SD
Cantaloupe/honeydew	80%	90%	350.9	CA, WI, MN, WA
Other fruits & nuts[c]	10% - 90%	10% - 90%	1,633.4	—
Other vegetables/melons[d]	70% - 100%	10% - 90%	1,099.2	—
Other field crops[e]	10% - 100%	20% - 90%	70.4	—
Total	—	—	$14,563.6	—

Source: Compiled by CRS using values reported in Morse, R.A., and N.W. Calderone, *The Value of Honey Bees as Pollinators of U.S. Crops in 2000*, March 2000, Cornell University, at [http://www.masterbeekeeper.org/pdf/pollination.pdf].

Figure 6

Honeybees are responsible for pollinating roughly 100 different kinds of fruits and vegetables produced in the United States (see Figure 6; Stokstad 2007). Dairy and beef products are also reliant on honeybee pollination because bees help facilitate large yields of alfalfa. Without honeybees, we would have a much more difficult time obtaining the foods that we consume everyday.

In the last half century, there has been a major expansion in the cultivation of pollinator-dependent crops due to modern industrialized agriculture, resulting in a high demand for pollination services. However, populations of honeybees, both managed and feral, have been steadily falling due to a variety of stress factors. These losses are directly affecting commercial agriculture industries. The almond industry, for example, is an important and lucrative business that exemplifies a crop heavily reliant on honeybee pollination. The United States supplies 80% of the world's almonds, and in 2006, almond exports were estimated to be worth $1.5 billion (Schacker 2008). Pollination of thousands of almond acres in California's central valley requires the rental of honeybee colonies from all over the country. As honeybee populations continue to dwindle and as almond acreage continues to expand, the price of renting colonies has increased.

From 2004 to 2006, the price of honeybees for almond pollination rose from $54 per colony to $136 per colony, directly affecting the cost of almonds (Sumner 2006). Also, more land acreage has been dedicated to growing almonds. In 1996, there were 430 thousand acres for almonds in California and in 2004, this number rose to 550 thousand and is expected to increase even more (Sumner 2006). Decreased populations of honeybees as well as the increased competition between industries needing pollinators during the same months of the year for their respective crop has become extremely problematic, especially for the producers and the beekeepers involved in commercial crop pollination.

Increased transportation of hives and poor nutrition has been known to cause increased problems for bees. Honeybee colonies are often loaded onto eighteen-wheel flatbed trucks for days and are shipped through a range of time zones, causing their immune systems to be compromised (Stokstad 2007). The number of U.S. honeybee colonies dropped from 5 million in 1940 to 2 million in 1989 and it has been suggested that economic shifts in farming are to blame (Stokstad 2007). Colony loss, however, is a complex issue. The varroa mite, for example, was introduced to the U.S. in 1987 and since the parasite has nearly eradicated all wild honeybee colonies, farmers have to rely on rented colonies (vanEngelsdorp et al. 2009). This has resulted in the growth of large-scale beekeeping operations, which is causing bees to be more at risk for disease. There is an unsustainable positive feedback loop that is occurring and as honeybees are affected by increased stress associated with industrialized agriculture, colony collapse continues to worsen.

Malnutrition

In the United States, hives that are rented for commercial purposes tend to move from the West in the spring, often for the almond crop, to the North, Midwest, and East in the summer for a variety of other crops, including blueberries (Schacker 2008). It is an intense process that is cumulatively putting a strain on the honeybee's immune system. Nonetheless, commercial pollination remains imperative for big agricultural growers and the beekeepers who count on the fixed income from this operation.

Malnutrition is a consequence of shipping bees cross-country to pollinate a crop. According to Eric Mussen, "Most bees used for commercial pollination are placed in areas where only one, or perhaps just a few, pollens are available" (Mussen 2011). He continued to say that, "A good mix of pollens is required to rear healthy bees and many times this is not the case" (Mussen 2011). Also, many commercial beekeepers who transport their bees long distances to pollinate crops use high-fructose corn syrup and soy protein to feed their bees during their travel. These food sources, however, are not the best replacements for the enzyme and nutrient-rich raw honey and pollen that normally make up a bee's diet (Schacker 2008). Mussen continued to say, "Feeding bees man-made pollen substitutes and supplements will increase bee numbers, but those bees are not as nutritionally well fed and physiologically robust as are bees that have been living on a mixed pollen diet" (Mussen 2011).

The transportation of hives throughout the country has been happening for years, even before CCD was reported. On the other hand, colony collapse has worsened since there have been increases in habitat loss and decreases in nectar and pollen biodiversity (Naug 2009). A loss of bee forage may be synergistically combining with disease as well as a variety of other stress factors to lower bee population. Further research needs to be conducted on the health of honeybees, especially at times when they are not receiving a diverse diet.

Genetic Loss

Scientists have been researching the lack of genetic diversity in managed honeybee colonies as another factor contributing to colony collapse disorder. The commercial beekeeping industry relies on only about 500 breeder queens to produce the millions of queens used to start colonies, which can be seen as a "genetic bottleneck" (Ellis 2009). The shortage of genetic diversity may be causing honeybees to become more susceptible to disease, despite the fact that honeybees have numerous defenses against parasites and pathogens (Oldroyd 2007). According to a survey that reported reasons of colony loss from 305 beekeeping operations in the U.S. (13.3% of managed colonies in the country), although starvation, varroa mites, and CCD were significant to colony loss, the primary problem for beekeepers was "poor queens" (vanEngelsdorp 2008).

In many cases, honeybee colonies are able to overcome times of infection because worker bees have an innate behavior to constantly rid the hive of diseased brood. However, for a colony to be resilient to pathogens and to overcome times of infection, it requires a high level of genetic variation (Oldroyd 2007). As the single egg layer in the hive, the queen's health it is critical. During mating flights, a queen will mate with an average of 12 drones, which is among the highest levels of polyandry in social insects, and this genetic variability is reflected in the gene base of her workers (Tarpy 2003).

Multiple studies have tested the association between genetic diversity and disease susceptibility in honeybees. Results have shown that the higher the genetic diversity within a hive, the more resilient the hive is to parasites and pathogens. This is because the worker bees have an increased fitness that allows them to engage in hygienic behavior more effectively and there is higher brood viability (Tarpy 2003). As artificial insemination of queens and honeybee domestication become more common and as the honeybee gene pool becomes smaller, infestations of parasites and pathogens will become more common. It is therefore becoming increasingly necessary to understand the factors that are contributing to "low-quality queens" (Delaney et al. 2010).

Pesticides

Many people believe that pesticides, especially a newly released class of insecticides, called neonicotinoids, are the cause of CCD (Kaplan 2008). Modern industrialized agriculture relies on vast amounts of chemicals to produce high crop yields. Also, as mites and viruses have become more prevalent, beekeepers are increasing how often they use miticides and chemicals to treat their colonies. Therefore, honeybees have become more exposed to pesticides when foraging as well as during times of infection.

Neonicotinoids

Beekeepers and scientists all over the world have become more interested in the harmful effects of pesticides, particularly since a new class of neuro-active insecticides, called neonicotinoids (modeled after nicotine) have been released on the market (Matsuda 2001). Neuro-active insecticides are extremely dangerous to honeybees because they disturb the organism's neurobehavioral and immune system, both of which are crucial to the insect's well being (Schacker 2008).

Most of today's farmers depend on chemicals for an increased yield. In the past, insecticides were sprayed aerially, but now most seeds are chemically treated before they are even planted. For example, imidacloprid, a neonicotinoid, is either painted on seeds or is poured around a plant in a "soil drench" (Schacker 2008). This is problematic because the toxins are now systemic, meaning that they are integrated throughout the entire plant. Thus, the toxins are likely to be taken up by honeybees even though they are not meant for the pollinators (Theobald 2010).

Corn, for example, is considered an important protein source for honeybees and covers more than 88 million acres of U.S. farmlands (Schacker 2008). Most conventional corn seeds and nearly all genetically modified corn seeds containing *Bt* are coated with nicotinyl seed treatments, making it nearly impossible for honeybees to resist exposure (Benbrook 2008). The two examples of neonicotinoids that will be discussed in this paper are imidacloprid and clothianidin, each of which is gaining momentum as causal agents of CCD (Theobald 2010).

Imidacloprid

Imidacloprid (IMD) is a neonicotinoid manufactured by Bayer CropScience under the name GAUCHO. It is a patented chemical used for pest control, seed treatments, and insecticide spray. IMD was first registered in the U.S. in 1994 and is now used on a wide variety of crops (Theobald 2010). IMD is similar to DDT in the sense that they are both neurotoxins and have properties similar to nerve gas and are both designed to block important parts of the pest's nervous system to keep them from properly functioning (Matsuda 2001). However, even though honeybees are not the targeted insect, they are experiencing the same symptoms. When the chemical was first authorized, Bayer's studies reported that GAUCHO was safe for bees because only the roots of the plant would get IMD and not the flower or the nectar (Schacker 2008). However, researchers have continued to conduct studies on neurotoxins and many of them have found that imidacloprid at a concentration of 100 ppb disrupts honeybee communication, therefore resulting in a decline in foraging activity (Bortolotti et al. 2003). Further research needs to be conducted in order to better portray the effects of imidacloprid on honeybees.

Clothianidin

Clothianidin is another example of a highly toxic neonicotinoid that has been said to be a culprit for honeybee deaths (see Figure 7). The pesticide is manufactured by Bayer CropScience under the name PONCHO and was granted registration by the EPA in 2003 (Mogerman 2008). Clothianidin is used to coat corn, soy, sugar beets, sunflowers, as well as other seeds and is now extremely common in the market (Kay 2008). Since its release, farmers have purchased more than $262 million worth of the insecticide (Keim 2010). Before a pesticide can be approved for use, a company must

Neonicotinoids' Toxicity to honey bees

Chemical	Brand name	Acute Contact	Acute Oral
thiamethoxam	Actara, Platinum, Helix, Cruiser, Adage, Meridian, Centric, Flagship	Highly toxic	Highly toxic
clothianidin	Poncho, Titan, Clutch, Belay, Arena	Highly toxic	Highly toxic
imidacloprid	Confidor, Merit, Admire, Ledgend, Pravado, Encore, Goucho, Premise	Highly Toxic	Highly toxic
acetamiprid	Assail, Intruder, Adjust	Toxic	Toxic
thiacloprid	Calypso	Toxic	Toxic
dinotefuran	Venom	Highly Toxic	Highly Toxic

Figure 7

submit a report attesting its safety to honeybees, and Bayer's report indicated that clothianidin was harmless to bees (Kay 2008). However, in 2008, a leaked EPA memo revealed flawed testing that Bayer conducted on honeybees. Later investigations reported that the study by Bayer was an unsuitable way to observe the effects of the pesticide on honeybees (Theobald 2010).

According to recent reports, the Bayer safety tests on clothianidin were conducted on two acre-plots of land for each of the treated crops. Honeybees, however, usually fly at least two miles away from the hive to forage, so most likely, the bees did not collect very much pollen from the treated crops. Also, the tests used treated canola rather than corn – the major plant species that bees rely on as a primary protein source in the winter (Theobald 2010). It appeared that Bayer designed tests that were not realistic with respect to known honeybee foraging patterns.

In a statement by commercial beekeeper Dave Hackenberg:

> What folks need to understand is that the beekeeping industry, which is responsible for a third of the food we all eat, is at a critical threshold for economic reasons and reasons to do with bee population dynamics. Our bees are living for 30 days instead of 42, nursing bees are having to forage because there aren't enough foragers and at a certain point a colony just doesn't have the critical mass to keep going...
>
> The bees are at that point, and we are at that point. We are losing our livelihoods at a time when there just isn't other work. Another winter of more studies are needed so Bayer can keep their blockbuster products on the market and EPA can avoid a difficult decision, is unacceptable.

Clearly, the fight between beekeepers and pesticide companies is a controversial issue. At this point, 121 different pesticides have been found in bees, wax, and pollen (Benjamin 2010). According to Eric Mussen, "Many beekeepers are convinced that exposure to pesticides is causing the problem. While it is true that ALL pesticides are detrimental to the physiology of exposed honeybees, residue analyses have shown that apparently healthy colonies have as many or more pesticide residues than collapsing colonies" (Mussen 2011). Although specific chemicals are known to have adverse sublethal effects on honeybees, further research is important in order for the scientific community to have a consistent understanding of the relationship between pesticides and CCD.

Cell Phones

Cell phones were once mistakenly thought to contribute to CCD. This idea came about when two German scientists at Landau University conducted research to better understand the effects of radioactive waves on honeybees' ability to navigate (Kuhn 2003). However, the study did not use a mobile phone, but rather a cordless phone base (Schacker 2008). *The Independent*, a UK-based newspaper, published a story called, "Are Mobile Phones Wiping Out Our Bees?" The story reported that the scientists claimed that cell phones were interfering with the bees' ability to navigate and that bees refused to return to their hives when cell phones were placed nearby (Lean 2007). Similar articles were published internationally, causing people to automatically relate cell phones to CCD. It turned out that *The Independent's* reporters got their facts wrong considering cell phones were not even used in the experiment and the reporters had not even spoken to the German scientists (Schacker 2008) It is imperative that colony collapse disorder is not undermined due to faulty communication between researchers and the media.

SECTION III: Effects of Colony Collapse Disorder

"If the bee disappears from the surface of the earth, man would have no more than four years to live."
~Albert Einstein~

If colony collapse disorder continues to worsen, there will be costly economic, ecological, and social effects. Honeybees play an integral role in the environment, which affects the biological system as a whole. As the biological system breaks down, humans will face consequences that will severely impact their health and their livelihoods. It is mandatory to consider these repercussions when discussing the possible extinction of the honeybee.

Economic Effects

According to the CCD Steering Committee, honeybees are economically worth $15 billion to the U.S. economy though the enabled production of food (CCD Steering Committee 2007). If CCD continues to worsen in the United States, there will be a variety of direct and indirect economic costs. Direct costs include the loss of jobs and commercial farming sectors, while indirect costs include increased trade deficits and food prices.

With losses of 50% or more each year of honeybee colonies, commercial beekeeping is already struggling to remain in business (Schacker 2008). If losses continue, there will be a rise in the cost of managing bees causing the price of honeybee rentals to increase. This price increase will severely impact many agriculture industries, including the California almond growers who rely on honeybee colonies to support the $2 billion almond industry (Sumner 2006). As production costs rise and crops that depend on honeybee pollination decrease, farmers will be forced to convert to growing plants that do not depend on pollination or abandon their farms all together (Munawar et al. 2009). Small farmers will be predominately affected because of the high variety of crops they grow and the lack of subsidies from the government. Also, dairy and beef farmers will be affected because their livestock depends on clover and other foraging crops for food (Berenbaum 2007).

As colony collapse disorder leads to larger local extinctions of honeybees, there will be a loss of crops that rely on pollination. This will lead to the importation of bees and food from foreign countries (Berenbaum 2007). If the United States is forced to import more food, the prices may be higher than they are now and the U.S. will face an increased trade deficit (Berenbaum 2007). This deficit will be compounded as production and commodity exports decrease. For example, the U.S. supplies 80% of the world's almonds, a crop that is heavily reliant on honeybees for pollination. If there is a significant loss of bees, then almond orchards face the risk of going bankrupt (Schacker 2008). Furthermore, the remaining beekeepers in the honey industry may be forced to leave the sector because they will be competing with cheaper, imported honey (Aizen 2009). There is also a major risk of importing bees to local fauna because of the introduction of new pests.

With increased pollination costs, the total price of crops grown in the United States will most likely increase and result in a reduction of consumer welfare (Berenbaum 2007). For example, if apple orchards on the Western Slope in Colorado are forced to import honeybee colonies for pollination (a service already in high demand), then there will be a rise in internal production costs – this increased cost will be passed on to consumers.

Pollination services support an entire system of agriculture, and if we do not take significant action, repercussions of further honeybee loss will distress the entire nation. If economists need to monetize nature to show the benefits of honeybee services to make positive environmental change, then I think it is well worth a more in-depth analysis to create agricultural and conservation policies.

Ecological Effects

As both a keystone species and index species, honeybees provide a service that supports a large range of ecosystems, while also indicating the health of their surrounding environment. Honeybees promote ecosystem vitality through services that provide pollination and help maintain genetic diversity by enabling reproduction of a wide variety of plant species. Biodiversity is extremely important for an ecological system because it allows for competition and natural selection among different plant species, thus reducing the risk of disease and vulnerability to pathogens. These plants also provide shelter for animals and generate the fruit and seeds on which they depend (Berenbaum 2007). If vital pollinators such as honeybees are lost, the ecological effects will be detrimental to an entire biological system.

Honeybees also serve as an index species because they are able to show the condition of the environment. The fact that honeybees are dying in significant numbers suggests that there is an important imbalance to be addressed in the environment. The analysis of honey, wax, and the bees themselves, is helpful in providing a better understanding of the diffusion of pesticides in the environment (Chauzat 2006). Studies have also indicated that radionuclide contamination and heavy metal contamination can also be monitored using honeybees (Chauzat 2006). The close relationship that these insects have with the environment allows them to serve as indicators for contamination. This is an important service to humans because, similar to bees, we require a healthy environment to remain a viable species.

Social Effects

If colony collapse continues to worsen and crops that depend on pollination become increasingly scarce, there will be dramatic social effects. It is possible that middle and low class families will be forced to pay three to four times more than what they already pay for food (Schacker 2008). After years of inflation and rising prices for oil, electricity, and healthcare, these added food costs will be significant and could potentially limit access to nutritional foods. If families do not receive nutritional food, health problems such as obesity and diabetes will most likely increase and further raise the price of healthcare (Schacker 2008).

In addition to the stratified effects described above, CCD will also affect the livelihoods of beekeepers and farmers. Generations of families who have farmed their land for centuries will be forced out of business. Effects will also be felt throughout the food sector as a whole. Grocery stores and restaurants will experience increased prices, scarcity of crops for food, and a diminished consumer base (Berenbaum 2007). With 11.2 million people working in the food service industry and 400,000 people working in the agricultural industry in the United States, a loss of honeybees will have significant repercussions (Bureau of Labor Statistics 2010).

Honeybees are responsible for pollinating a diverse range of flowering plants, thus maintaining ecological diversity and providing valuable commodities for humans. There will be severe losses ecologically, agriculturally, and economically if honeybee populations cease to exist. It is mandatory for people to evaluate the effects of continued colony loss as an incentive to promote honeybee awareness and contribute to preventing further losses.

SECTION IV: Mitigation Strategies for Colony Collapse Disorder

"We allow the chemical death rain to fall as though there were no alternatives, whereas in fact there are many, and our ingenuity could soon discover many more if given opportunity."

~Rachel Carson~

Policy Recommendations

The consequences of colony collapse disorder are broadly distributed among the general public. Moreover, the potential causes of CCD - such as pesticides, commercial pollination, and decreasing biodiversity in agriculture - are firmly ingrained into the daily processes of our society. Therefore, government policies must play an important role in the protection of public health and the promotion of a safer environment for key pollinators.

Examples of useful policies include: pesticide bans, rooftop beekeeping in urban environments, funding and technical support for community gardening, organic farming, public education, scientific research, and support for local and commercial beekeepers. Several countries that have also been impacted by colony collapse have reacted to the disorder through a mix of public education initiatives, policies, and further research. Some of these efforts are described below, in addition to examples of local government action in the United States.

France

In July of 1994, French beekeepers noticed that shortly after sunflowers were in full bloom, hives began to collapse (Theobald 2010). Worker bees flew off and never returned, leaving the queen and some brood to die. Beekeepers also noted that

GAUCHO, a new insecticide containing imidacloprid (IMD), was recently approved as a product by Bayer CropScience (Schacker 2008). Between 1996 and 1999, production of sunflower honey fell from 110,000 metric tons to 50,000 tons and between 1995 and 2000, 76% of French apiaries suffered high winter losses of their bees (Schacker 2008). French beekeepers appealed to the government for help while French researchers began conducting a variety of studies (Underwood 2010). The protest by beekeepers was successful. In 1999, France banned IMD on sunflowers and in 2003 banned the chemical on sweet corn (Underwood 2010). However, even after the ban was put into place, French beekeepers continued to see declines in their bee populations. They discovered another insecticide in the soil, Regent, that contains fipronil (Schacker 2008). Fipronil is an insecticide that was commonly used to coat sunflower seeds in the South of France and tests showed that the insecticide negatively affects honeybee perception, olfactory learning, and motor function; all which are required for foraging behavior (Hassani et al. 2005). As a precautionary measure, in 2004, the head of ministry suspended the use of fipronil, and by 2005, beekeepers saw a return in their bee populations (Schacker 2008). The resurgence in honeybees suggested that IMD and fipronil were the culprits causing CCD. The fight in France to ban pesticides set a precedent for beekeepers and government agencies around the world: create policies to limit harmful environmental toxins in order to prevent further colony collapses.

Germany

In May of 2008, the Baden-Wurtemberg region of Germany lost two-thirds of its colonies (Theobald 2010). Scientists traced the losses to pesticides, specifically clothianidin, which had been registered in Germany in 2004 (Theobald 2010). Within two weeks, Germany suspended the registration of clothianidin (Schacker 2008). Since France and Germany have banned imidacloprid, fipronil, and clothianidin, many other countries and provinces have followed suit, including Italy, Japan, the United Kingdom, and British Columbia.

New York

In addition to pesticide bans, governments have engaged in other strategies to fight colony collapse disorder, including methods to encourage beekeeping and promote bee awareness. For example, in the spring of 2010, New York overturned its statewide anti-beekeeping ordinance (Navarro 2010). The original fear of policymakers was that bees were too dangerous to be in a busy city and that people would be stung. By making urban beekeeping legal, beekeepers were able to show that honeybees are not dangerous creatures and are actually beneficial. Andrew Cote, president of the New York City Beekeepers Association said, "Honeybees are interested in water, pollen, and nectar and the real danger is the skewed public perception of the danger of honeybees" (Navarro 2010). Beekeeping is still illegal in 89 U.S. cities, a policy that should be changed in order to promote the collective welfare of honeybees and humans (LeVaux 2010).

Other Recommendations and Strategies

There are a number of actions that our communities can take to reverse or eliminate colony collapse disorder. First, we can increase the funding that we put into research on this pressing issue. It is important to collect data on pollination services, colony movement between states, and the general health of bees. Such information will show the economic value of honeybees as well as their condition (Berenbaum 2007). Also, it is important to better examine wild bees. Wild pollinators are important for crop pollination and are actually more versatile than the honeybees, but are also experiencing rapid loss (Berenbaum 2007). As honeybees continue to disappear, there needs to be a better understanding of the impacts that other native pollinators have on our agricultural system and larger ecosystem, as well as their current level of vitality. This information will be useful as researchers and elected officials begin to implement strategies to cope with the effects of CCD in the effort to sustain agriculture over time.

It is also important for elected officials and community leaders to engage in actions that promote sustainable agriculture, foster bee awareness, and help to reverse CCD. For example, Michelle Obama made a statement when she announced the White House was going to plant an organic vegetable garden as a way to promote healthy living and the need to tackle obesity  throughout the country. The garden also includes a beehive that helps to pollinate crops and serve the educational mission of the program (see Figure 8, Flottum 2009). Planting gardens and keeping bees are some ways that the general public can better appreciate the importance of honeybees and the direct role that these pollinators play in our lives. By being directly involved with bees, people can better understand colony collapse disorder and make informed decisions about honeybee policies.

Many people who live in cities and do not have access to garden space or room have begun keeping beehives on their roofs. Policies have been enacted that allow people to keep hives on the roofs of apartments and municipal buildings (see Figure 9). Mayor Richard Daley of Chicago put beehives on top of City Hall (Hinke 2007). Harvesting of the honey is done by a trained beekeeper in conjunction with city youth programs. The 200 pounds of honey that they retrieve annually is jarred and sold in the city hall gift shop, and the proceeds

Figure 9

benefit Chicago Cultural Center projects (Hinke 2007). The Chicago urban beehive project is now a model for other cities interested in expanding the welfare of bees. This is a perfect example of how to tie the importance of honeybees into an entire city. Urban hives are a way to promote bee awareness and support healthy bee colonies.

Education is another important strategy to combating CCD. Colleges and communities can both have a major impact. It is important to educate their constituents about the stressors that are playing a role in honeybee disappearance and to provide them with opportunities that will contribute to the prevention of further losses. Personally, I have been involved with a number of efforts to promote education in both my college and my community. At the University of Colorado at Boulder, I planned an event to screen the movie, *Vanishing of the Bees*. This film addresses the broad spectrum of what may be contributing to colony collapse as well as provides ways for viewers to take action. In addition to public education events, CU has pledged to completely rid the campus of pesticides by 2015. As described earlier in this paper, pesticides have been demonstrated to have severe consequences to honeybees. As part of my strategy to combat colony collapse, I contributed a crucial element to the campus pesticide reduction plan; more specifically, a brief excerpt on colony collapse disorder and a description of how honeybees are affected by the widespread use of pesticides (see Appendix 3). I hope that this plan to make the campus pesticide-free can serve as a model for other universities.

In my Boulder community, I have joined a group of beekeepers and concerned citizens. The group, called *Coalition 4 B's* (bees, birds, bats, butterflies), has bi-monthly meetings to discuss ways in which we can promote honeybee awareness and plan events where we can educate the general public. We have also worked with other groups around the United States to combat honeybee loss. Contributions such as these are some of the most effective ways ordinary citizens can help resolve the problem of honeybee losses. Through grassroots organization and willingness to change, communities can influence the decisions of elected officials and encourage them to pass strong policies that support the health of our most vital pollinator.

Conclusion

Since CCD was reported in 2006, beekeepers have seen as many as 90% of the bees within their hive disappear, making this issue of great importance (Mogerman 2008). In testimony before Congress, May Berenbaum said, "If honeybee numbers continue to decline at rates documented from 1989 to 1996, managed honeybees will cease to exist by 2035" (Berenbaum 2007). Honeybee pollination is an essential service that must be preserved. Honeybees need to be given an economic valuation in order for their services to be quantified (Draxler 2011). Right now, they are worth $15 billion to American agriculture, but their ability to support an entire ecosystem is worth much more. If there are no more honeybees to pollinate crops, the food supply would not be completely destroyed in terms of how many calories we consume. Plants that do not require insect pollination such as grains could still be produced, but the bulk of the food we consume for vitamins and nutritional content would not exist if there were no insects to pollinate them (Berenbaum 2007). Food security is at risk, especially as our population continues to rise.

A perfect storm of stressors is putting the honeybees at risk of extinction. The lack of diversity within our agricultural system is even more dangerous because there is not enough variety in agro-ecosystems to support healthy pollinator communities (Steffan-Dewenter 2005). If we do not change how crops are grown and the amount of chemicals that are dumped into the environment every day, pollinators will continue to suffer and humans will risk losing one of our most valued necessities, food.

In the winter of 2009 alone, according to the Apiary Inspectors of America and the US government's Agricultural Research Service, the number of managed honeybee colonies in the United States fell by 33.8% (Benjamin 2010). The pollen collecting skills that a honeybee possesses are vital for the lives of farmers, the environment, and humanity as a whole that depend on crops to survive. Unfortunately, there has still been a devastatingly slow response by the federal government.

Further Research

The 30% loss of honeybees suffered in the last several years to colony collapse disorder has prompted ecologists to investigate other pollinators. All pollinators depend on a healthy ecosystem. Already, bats, hummingbirds, and butterflies have been disappearing (Schacker 2008). Although I examined many of the current hypotheses surrounding colony collapse, other factors are also being considered. These include the effects of habitat loss and the overall effects of climate change (Konkel 2009).

Habitat loss is most likely not a direct cause to colony collapse disorder, but the loss of open space for foraging is putting pollinators at a major risk. Habitat is being lost because of agriculture, grazing, fragmentation, and development. Honeybees rely on a variety of flowers and flowering plants for nectar and pollen sources and as fragmentation increases, pollinator species and numbers are decreasing (Kearns 1997). When there are fewer pollinators, plant reproduction and fitness are adversely affected. Also, when there are fewer pollinators, the production of seeds and fruits is reduced, which affects all the species that depend on these sources of food. With entire ecosystems dependent on pollinators for survival, it is wise to include habitat loss in the discussion concerning CCD.

In the past century, large-scale monocultures have been the most prevalent way to farm and by doing this, the amount of wild vegetation has been compromised. This affects honeybees because it is important for them to be able to feed from a wide variety of sources for pollen and nectar (Kearns 1997). Also, increased development is causing more habitat fragmentation, which also may affect the bees and other pollinating species that forage in a specific way. For example, flowers that are in a small population may be passed by pollinators that usually look for large patches of flowers, thus causing certain plants to not get mandatory pollination services (Kearns 1997). Biologists who study habitat fragmentation speculate the possible loss of species because of the ecological changes that result from it. They have also said that the collapse of pollinator-plant relationships is likely as well, but further studies need to be done (Harrison 1999).

Another factor that has shown to be correlated with CCD is climate change. Weather patterns affect bee's foraging behavior. As warmer temperatures and the severity of droughts become more prevalent, there could be major effects on pollinators. Droughts will limit the amount of available water causing bees to be at increased risk for fatality (Willmer 1991). Also, oftentimes-warm weather allows the increased migration of parasites and pathogens. The spreading of disease is already a major problem facing honeybees and climate change could worsen the situation. As global temperatures are steadily rising, it is important to understand the potential effects in order to take proactive measures to save this vital species.

Although further research needs to be conducted on other potential causes that may be contributing to colony loss, it can be said with certainty that the amount of disease infecting honeybees rose as the number of toxic chemicals available to treat them increased and modern beekeeping turned into a business to produce the maximum profit with little regard to the honeybee's overall health. Natural processes to treat honeybee infections have been slowly disregarded and organic beekeeping has become nearly impossible to do because of the widespread use of pesticides. Honeybees are being pushed past their normal work capacity and becoming stressed by a variety of factors. These incredible social creatures have been functioning as a system for millions of years and the more fast-paced that human society becomes and the more rapidly we expect these bees to work for us, the more collapse we will see.

Bibliography

Aizen, Marcelo A., and Lawrence D. Harder. "The Global Stock of Domesticated Honey Bees Is Growing Slower Than Agricultural Demand for Pollination." *Current Biology* 19.11 (2009): 915-18.

Ambrose, John T. (2000). *Varroa Mite Disease*. Raleigh: N.C. State University, Department of Entomology and North Carolina Cooperative Extension.

Antúnez, Karina, Matilde Anido, Geraldine Schlapp, Jay D. Evana, and Paplo Zunino. "Characterization of Secreted Proteases of Paenibacillus Larvae, Potential Virulence Factors Involved in Honeybee Larval Infection." *Journal of Invertebrate Pathology* 102 (2009): 129-32.

Benbrook, Chuck. "Prevention, Not Profit, Should Drive Pest Management." *Pesticides News* 82 (2008). The Organic Center, Dec. 2008. Web. 5 Apr. 2011. http://www.organic-center.org.

Benjamin, Alison. "Pesticides: Germany Bans Chemicals Linked to Honeybee Devastation." *Latest News, Comment and Reviews from the Guardian.* 23 May 2008. Web. 22 Feb. 2011. http://www.guardian.co.uk.

Benjamin, Alison, and Brian McCallum. *A World without Bees*. New York: Pegasus, 2009.

Berry, Jennifer, ed. "CCD Finally Revealed? Israeli Acute Paralysis Virus (IAPV) Linked to CCD Colonies." *Georgia Bee Letter* 18.3 (2007): 1-13.

Bortolotti, Laura, Rebecca Montanari, Jose Marcelino, Piotr Medrzycki, Stefano Maini, and Clausio Porrini. "Effects of Sub-lethal Imidacloprid Doses on the Homing Rate and Foraging Activity of Honey Bees." *Bulletin of Insectology* 56.1 (2003): 1721-8861.

Caron, Dewey M. (2000). *Trachael Mites*. Mid-Atlantic Research and Extension Consortium. U.S. Department of Agriculture. 4.2, 1-3.

CCD Steering Committee. *Colony Collapse Action Plan*. ARS USDA Executive Summery. 20, June 2007. Web. 06 Dec. 2010.

Chauzat, Marie-Pierre. "A Survey of Pesticide Residues in Pollen Loads Collected by Honey Bees in France." *Journal of Economic Entomology* 99.2 (2006): 253-62.

Colony Collapse Disorder and Pollinator Decline, 110th Cong. (2007) (testimony of May R. Berenbaum).

Conrad, Ross. "American Foulbrood-A Review." *Bee Culture, 1.1* (2009): 54-56.

Conrad, Ross (2009). Colony collapse disorder: CCD - the sign of things to come? *Bee Culture, 137*(6), 47-48.

Crane, Eva. "A Short History of Knowledge about Honey Bees (Apis) up to 1800." *Bee World* 85.1 (2004): 6-11. *International Bee Research Association*. Mar. 2004. Web. 6 Dec. 2010. http://www.ibra.org.uk.

Debnam, Scott. "Colony Collapse Disorder: The Symptoms Change with the Seasons, and Are Different in Various Locations." *Bee Culture* (2009): 30-32.

Delaney, Deborah A., Jennifer J. Keller, Joel R. Caren, and David R. Tarpy. "The Physical Insemination, and Reproductive Quality of Honey Bee Queens (Apis Mellifera L.)." *Apidologie* (2010): 1-13.

Delaplane, Keith S., and D. F. Mayer. *Crop Pollination by Bees*. Wallingford, England: CABI, 2000.

Delaplane, Keith S. "Africanized Honeybees." *University of Georgia Cooperative Extension Bulletin* 1290 (2010): 1-4.

Donzé, Gérard, and Patrick M. Guerin. "Behavioral Attributes and Parental Care of Varroa Mites Parasitizing Honeybee Brood." *Behavioral Ecology and Sociobiology* 34.5 (1994): 305-19.

Draxler, Breanna. "Does Placing a Price Tag on Natural Resources Make Them More Valuable? | Guest Writer | Travel & Outdoors | NewWest.Net." *NewWest.Net | Colorado, Idaho, Montana, New Mexico, Utah, Wyoming*. 22 Feb. 2011. Web. 11 Mar. 2011. http://www.newwest.net.

Ellis, Jamie. "Colony Collapse Disorder (CCD) in Honey Bees." *EDIS*. University of Florida IFAS Extension, 2009. Web. 15 Mar. 2011. http://edis.ifas.ufl.edu.

Flinn, Pat. E-mail interview. 09 Feb. 2011.

Flores, J.M., I. Gutierrez, and R. Espejo. "The Role of Pollen in Chalkbrood Disease in Apis Mellifera: Transmission and Predisposing Conditions." *Mycologia* 97.6 (2005): 1171-176.

Flottum, Kim. "An Addition to Michelle Obama's White House Garden: Honey Bees." *The Daily Green*. 23 Mar.2009. Web. 08 Mar. 2011. http://thedailygreen.com.

Genersch, Elke, Jay D. Evans, and Ingemar Fries. "Honey Bee Disease Overview." *Journal of Invertebrate Pathology* 103 (2010): 2-4.

Gilliam, Martha, Stephen Taber, and Gary V. Richardson. "Hygienic Behavior Of Honey Bees In Relation To Chalkbrood Disease." *Apidologie* 14.1 (1983): 29-39.

Harrison, Susan, and Emilio Bruna. "Habitat Fragmentation and Large-scale Conservation: What Do We Know for Sure?" *Ecography* 22.3 (1999): 225-32.

Hassani, Abdessalam K., Matthieu Dacher, Monique Gauthier, and Catherine Armengaud. "Effects of Sublethal Doses of Fipronil on the Behavior of the Honeybee (Apis Mellifera)." *Elsevier* 82 (2005): 30-39.

Hinke, Veronica. "Here & There: Beekeeping on a Chicago Rooftop Garden." *Herb Companion Magazine*. Jan. 2007. Web. 08 Mar. 2011. http://www.herbcompanion.com.

Horn, Tammy. "Honey Bees: A History - NYTimes.com." *Topics - Times Topics Blog - NYTimes.com*. 11 Apr. 2008. Web. 06 Dec. 2010. http://topics.blogs.nytimes.com.

Johnson, Renee. "Recent Honey Bee Colony Declines." *Congressional Research Service*. 26 Jan. 2009. Web. 6 Dec. 2010. <http://www.crs.gov>.

Kaplan, J. K. "Colony Collapse Disorder: A Complex Buzz." *Agricultural Research* 56.5 (2008): 8-11.

Kay, Jane. "Lawsuit Seeks EPA Pesticide Data." *San Francisco Bay Area — News, Sports, Business, Entertainment, Classifieds: SFGate*. 19 Aug. 2008. Web. 13 Jan. 2011. http://www.sfgate.com.

Kearns, Carol A., and David W. Inouye. "Pollinators, Flowering Plants, and Conservation Biology." *BioScience* 47.5 (1997): 297-307.

Keim, Brandon. "Leaked Memo Shows EPA Doubts About Bee-Killing Pesticide | Wired Science | Wired.com." *Wired.com*. 13 Dec. 2010. Web. 13 Jan. 2011. http://www.wired.com.

Kolmes, S. A., M. L. Winston, and L. A. Fergusson. "The Division of Labor among Worker Honey Bees (Hymenoptera: Apidae): The Effects of Multiple Patrilines." *Journal of the Kansas Entomological Society* 62.1 (1989): 80-95.

Konkel, Lindsey. "Buzzing About the Future of Pollinators » Scienceline." *Scienceline*. 29 Jan. 2009. Web. 11 Mar. 2011. http://scienceline.org.

Kremen, Claire, Neal M. Williams, and Robbin W. Thorp (2009). *Crop Pollination from Native Bees at Risk from Agricultural Intensification*. 99.26, 16812-6816. Proceedings of the National Academy of Sciences of the United States of America. http://www.pnas.org.

Kuhn, Jochen, and Hermann Stever. "How Electromagnetic Exposure Can Influence Learning Processes – Modelling Effects of Electromagnetic Exposure on Learning Processes." University of Koblenz-Landau, Campus Landau, 2003. Web. 4 Apr. 2011. http://abeillesperdues.ouvaton.org.

Lean, Geoffrey, and Harriet Shawcross. "Are Mobile Phones Wiping out Our Bees? - Nature, Environment - The Independent." *The Independent | News | UK and Worldwide News | Newspaper*. 15 Apr. 2007. Web. 04 Apr. 2011. http://www.independent.co.uk.

LeVaux, Ari. "Resilient Ideas: Rooftop Beekeeping." *Yes!* Positive Futures Network, 9 Sept. 2010. Web. 8 Mar. 2011. http://www.yesmagazine.org.

Matsuda, Kazuhiko. "Neonicotinoids: Insecticides Acting on Insect Nicotinic Acetylcholine Receptors." *Trends in Pharmacological Sciences* 22.11 (2001): 573-80.

Meyerowitz, Steve. "Will You Bee Mine?" *Better Nutrition* 62.5 (2000): 54-58.

McGregor, Samuel Emmett. *Insect Pollination of Cultivated Crop Plants*. Washington: Agricultural Research Service, U.S. Dept. of Agriculture, 1976.

Mogerman, Josh. "NRDC: Press Release - EPA Buzz Kill: Is the Agency Hiding Colony Collapse Disorder Information?" *NRDC: Natural Resources Defense Council - The Earth's Best Defense*. 18 Aug. 2008. Web. 13 Jan. 2011. http://www.nrdc.org.

Munawar, Muhammad S., Shazia Raja, Mahjabeen Siddique, Shahid Niaz, and Muhammad Amjad. "The Pollination by Honeybee (Apis Mellifera L.) Increases Yield of Canola (Brassica Napus L.)." *Pakistan Entomology* 31.2 (2009):103-06.

Mussen, Eric. E-mail interview. 26 Jan. 2011.

Naug, Dhruba. "Nutritional Stress Due to Habitat Loss May Explain Recent Honeybee Colony Collapses." *Biological Conservation* 142.10 (2009): 2369-372.

Navarro, Mireya. "Bees in the City? New York May Let the Hives Come Out of Hiding." *The New York Times*. 14 Mar. 2010. Web. 8 Mar. 2011. http://www.nytimes.com.

Neumann, Peter, and Norman L. Carrack. "Honey Bee Colony Losses." *Journal of Apicultural Research* 49.1 (2010): 1-6.

Palacios, G., J. Hui, P. L. Quan, A. Kalkstein, K. S. Honkavuori, A. V. Bussetti, S. Conlan, J. Evans, Y. P. Chen, D. VanEngelsdorp, H. Efrat, J. Pettis, D. Cox-Foster, E. C. Holmes, T. Briese, and W. I. Lipkin. "Genetic Analysis of Israel Acute Paralysis Virus: Distinct Clusters Are Circulating in the United States." *Journal of Virology* 82.13 (2008): 6209-217.

Paxton, Robert J. "Does Infection by Nosema Ceranae Cause "Colony Collapse Disorder" in Honey Bees (Apis Mellifera)?" *Journal of Apicultural Research* 49.1 (2010): 80-84.

Pinto, M. Alice, William L. Rubink, John C. Patton, Robert N. Coulson, and J. Spencer Johnston. "Africanization in the United States: Replacement of Feral European Honey Bees (Apis Mellifera L.) by an African Hybrid Swarm." *Genetics* (2005): 1-93

Readicker-Henderson, Ed, and Ilona. *A Short History of the Honey Bee: Humans, Flowers, and Bees in the Eternal Chase for Honey*. Portland: Timber, 2009.

Roubik, David W. "Competitive Interactions between Neotropical Pollinators and Africanized Honeybees." *Science* 201.4360 (1978): 1030-032.

Sammataro, Diana, Uri Gerson, and Glen Needham. "Parasitic Mites of Honey Bees: Life History, Implications, and Impact." *Annual Review of Entomology* 45.1 (2000): 519-48.

Schacker, Michael. *A Spring without Bees: How Colony Collapse Disorder Has Endangered Our Food Supply*. Guilford, Connecticut: The Lyons Press, 2008.

Snodgrass, Robert E. *Anatomy of the Honey Bee*. Ithaca, NY: Comstock, 1984.

Somerville, Douglas, and Michael Hornitzky. "Nosema Disease." *Primefacts* 699 (2007): 1-3.

Spivak, Marla, and Gary S. Reuter. "Resistance to American Foulbrood Disease by Honey Bee Colonies Apis Mellifera Bred For Hygienic Behavior." *Apidologie* 32 (2001): 555-65.

Steffan-Dewenter, Ingolf, Simon G. Potts, and Laurence Packer. "Pollinator Diversity and Crop Pollination Services Are at Risk." *TRENDS in Ecology and Evolution* 20.12 (2005): 651-52.

Stokstad, E. "ENTOMOLOGY: The Case of the Empty Hives." *Science* 316.5827 (2007): 970-72.

Sumner, Daniel A., and Hayley Boriss. "Bee-conomics and the Leap in Pollination Fees." *Giannini Foundation of Agricultural Economics*. Department of Agriculture and Resource Economics at UC Davis, 14 Jan. 2006. Web. 4 Apr. 2011. http://aic.ucdavic.edu/research.

Tarpy, David R. "Genetic Diversity within Honeybee Colonies Prevents Severe Infections and Promotes Colony Growth." *Proceeding of the Royal Society B: Biological Sciences* 270.1510 (2003): 99-103.

Tarpy, David R., and Thomas D. Seeley. "Lower Disease Infections in Honeybee (Apis Mellifera) Colonies Headed by Polyandrous vs. Monandrous Queens." *Naturwissenschaften* 93 (2006): 195-99.

Tarpy, David R., Joshua Summers, and Jennifer J. Keller. "Comparison of Parasitic Mites in Russian-Hybrid and Italian Honey Bee (Hymenoptera: Apidae) Colonies across Three Different Locations in North Carolina." *Journal of Economic Entomology* 100.2 (2007): 258-66.

Tautz, Juergen. "Honeybee Waggle Dance: Recruitment Depends on the Dance Floor." *The Journal of Experimental Biology* 199 (1996): 1375-381.

Tentcheva, D., L. Gauthier, N. Zappulla, B. Dainat, F. Cousserans, M. E. Colin, and M. Bergoin. "Prevalence and Seasonal Variations of Six Bee Viruses in Apis Mellifera L. and Varroa Destructor Mite Populations in France." *Applied and Environmental Microbiology* 70.12 (2004): 7185-191.

"The Colony and Its Organization." *MAAREC - Mid Atlantic Apiculture & Extension Consortium*. 2011. Web. 24 Feb. 2011. https://agdev.anr.udel.edu.

"The Threat of Neonicotinoid Pesticides on Honeybees, Ecosystems, and Humans." *San Diego Beekeeping Society*. Japan Endocrine-disruptor Preventive Action, 27 Jan. 2011. Web. 22 Feb. 2011. http://www.kokumin-kaigi.org.

Theobald, Tom. "Do We Have a Pesticide Blowout?" *Bee Culture* (2010): 66-69.

Underwood, Robyn M. and Dennis vanEngelsdorp. "Colony Collapse Disorder: Have We Seen This Before?" *The Pennsylvania State University Department of Entomology*. Web. 12 Feb. 2010.

United States. Department of Agriculture. National Agricultural Statistics Service. *Honey*. Washington, D.C.: 2007.

United States. Department of Agriculture. Agricultural Research Service. *Colony Collapse Disorder: A Complex Buzz*. 2008.

United States. Department of Labor. Bureau of Labor Statistics. *Occupational Employment and Wages-May 2009*. Washington, D.C.: 2010.

VanEngelsdorp, Dennis, Jerry Hayes, Robyn M. Underwood, and Jeffery Pettis. "A Survey of Honey Bee Colony Losses in the U.S., Fall 2007 to Spring 2008." Ed. Nick Gay. *PLoS ONE* 3.12 (2008): 1-6.

VanEngelsdorp, Dennis, Jay D. Evans, Claude Saegerman, Chris Mullin, Eric Haubruge, Bach Kim Nguyen, Maryann Frazier, Jim Frazier, Diana Cox-Foster, Yanping Chen, Robyn Underwood, David R. Tarpy, and Jeffery S. Pettis. "Colony Collapse Disorder: A Descriptive Study." Ed. Justin Brown. *PLoS ONE* 4.8 (2009): 1-17.

Vivian, John. *Keeping Bees*. Charlotte, VT: Williamson Pub., 1986.

Williams, G. R., Tarpy, D. R., vanEngelsdorp, D., Chauzat, M.-P., Cox-Foster, D. L., Delaplane, K. S., Neumann, P., Pettis, J. S., Rogers, R. E. L. and Shutler, D. (2010), Colony Collapse Disorder in context. BioEssays, 32: 845–846.

Willmer, P.G. "Constraints on Foraging by Solitary Bees." *The Behavior and Physiology of Bees*. Melksham, UK: Redwood Press, 1991.

Winston, Mark L. *The Biology of the Honey Bee*. Cambridge, MA: Harvard UP, 1987.

Appendix 1

Possible causes of honeybee losses

Colony Collapse Disorder (CCD) ?

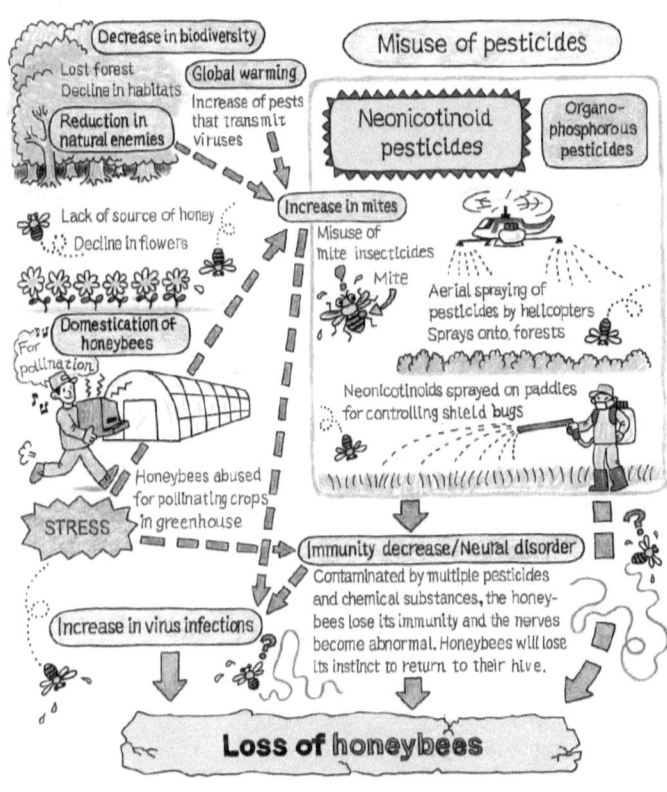

Illustration: Saori Yasutomi

Appendix 2

Avoid using these neonicotinoid pesticides
Synthetic nicotine-based pesticides toxic to honeybees and native pollinators

Ingredient names of products used in agriculture
Acetamiprid   ADJUST, ASSAIL, CHIPCO, INTRUDER, PRISTINE
Clothianidin   ARENA, BELAY, CLUTCH, PONCHO, TITAN
Dinotefuran   VENOM
Imidacloprid   ADMIRE (used on potatoes, corn, grapes, vegetables, citrus), ADVANTAGE,
    CONFIDOR, GAUCHO (used on corn, cotton, potatoes), HACHIUSAN, KOHINOR,
    LEVERAGE (cotton), MERIT (turf), PREMISE (termites), PROTHOR,
    PROVADO (fruits, vegetables), WINNER
Thiacloprid   CALYPSO (used on apple, pear, quince, crabapple)
Thiamethoxam ACTARA, ADAGE, CENTRIC, CRUISER, FLAGSHIP, HELIX, MERIDIAN, PLATINUM

We are asking growers not to use these products until more research is done.

Products found in local nurseries and hardware stores for home use:
AVOID:
Bonide Systemic Insect Spray: **Look for the active ingredient: Imidacloprid**
Bonide Systemic Insect Granules: *Active Ingredient*: Imidacloprid
Bonide Systemic Houseplant Insect Control: *Active Ingredient*: Imidacloprid
Bayer Season Long Grub Control: *Active Ingredient*: Imidacloprid
Bayer Advanced 3 in 1 Insect Disease and Mite Control: *Active Ingredient*: Imidacloprid
Bayer Advanced 2 in 1 Systemic Rose & Flower Care: *Active Ingredient*: Imidacloprid
Bayer Advanced 12 Month Tree & Shrub Protect & Feed: *Active Ingredient*: Imidacloprid
Bayer Advanced Tree & Shrub Insect Control 12 month: *Active Ingredient*: Imidacloprid
Bayer Advanced Dual Action Rose & Flower Insect Killer: *Active Ingredient*: Imidacloprid
Bayer Advanced Lawn Season Long Grub Control: *Active Ingredient*: Imidacloprid
Bayer Advanced Lawn Complete Insect Killer for Soil & Turf: *Active Ingredient*: Imidacloprid
Bayer Advanced Fruit Citrus & Vegetable Insect Control: *Active Ingredient*: Imidacloprid
Bayer Termite Control: *Active Ingredient*: Imidacloprid
Bayer All in One Rose and Flower Care: *Active Ingredient*: Imidacloprid
Ortho Max Tree & Shrub Insect Control: *Active Ingredient*: Imidacloprid
Ortho Max Flower, Fruit & Vegetable Insect Killer: **Look for the active ingredient Acetamiprid**
Ortho Rose Pride Insect Killer: *Active Ingredient* : Acetamiprid
Green Light Tree & Shrub Systemic Insect Killer: *Active Ingredient*: Imidacloprid
Green Light Systemic Rose & Flower Care: *Active Ingredient*: Imidacloprid

Never use a neonicotinoid pesticide on a blooming crop or on blooming weeds if honeybees are present.

Appendix 3

Excerpt from *The Turf Management Task Force Summary*

<u>Colony Collapse Disorder</u>

Colony collapse disorder (CCD), sometimes referred to as honeybee depopulation syndrome (HBPS), is a phenomenon in which worker bees from a beehive abruptly disappear. Symptoms include the rapid loss of adult worker bees, few or no dead bees found in the hive, and only a small cluster of bees with a live queen present and pollen and honey stores remaining in the hive. It was first reported in the United States in 2006 and has been dramatically affecting hives across the nation since. A direct cause has not yet been concluded for CCD, but potential stressors of this problem include commercial land-use, mites, pathogens, pesticides, and insecticides. A combination of these may be to blame, but studies have not yet found results that could be fully responsible for the problem, and may never, due the various concerns that could be causing the death of honeybees[1]. As a member of the Boulder community, the University should limit its negative impacts on honeybees that will significantly affect local beekeepers, farmers, gardeners, and the general public.

Pesticides have often been suspected as the cause of CCD, and many studies conducted by the USDA and the EPA as well as by governmental agencies abroad, like the French Agriculture Ministry, have noted that various pesticides affect a bee's ability to forage and may affect the fertility of a colony's queen. Additional research has indicated that pesticides can be lethal to honeybees. Sublethal pesticide effects, however, are subtler, although tests indicate that even small doses of pesticide exposure can affect honey production, cause foragers to disappear and kill off colonies (Underwood, vanEngelsdorp: 4). [2]

---

[1] VanEngelsdorp D, Evans JD, Saegerman C, Mullin C, Haubruge E, et al. (2009). Colony Collapse Disorder: A Descriptive Study. PLoS ONE 4(8): e6481. doi:10.1371/journal.pone.0006481.

[2] Underwood, Robyn M. and Dennis vanEngelsdorp. "Colony Collapse Disorder: Have We Seen This Before?" *The Pennsylvania State University Department of Entomology.* Web. 12 Feb. 2010. Print.

Appendix 4

December 8, 2010

The Honorable Lisa P. Jackson
Administrator
U.S. Environmental Protection Agency
Ariel Rios Building, MC 1101A 1200 Pennsylvania Avenue NW
Washington DC 20004

Dear Administrator Jackson:

In light of new revelations by your agency in a November 2, 2010 memorandum that a core registration study for the insecticide clothianidin has been downgraded to unacceptable for purposes of registration, we are writing to request that you take urgent action to stop the use of this toxic chemical. Clothianidin is a widely used pesticide linked to a severe and dangerous decline in pollinator populations. As we are sure you appreciate, the failure of the agency to provide adequate protection for pollinators under its pesticide registration program creates an emergency with imminent hazards: Food production, public health and the environment are all seriously threatened, and the collapse of the commercial honeybee-keeping industry would result in economic harm of the highest magnitude for U.S. agriculture.

The debate on clothianidin and the neonicotinoid pesticides is not new to the agency, but the recognition of the past failure of the Office of Pesticide Program's (OPP) 2007 scientific review, now acknowledged, requires immediate action to stop use while new studies are conducted. We refer you to the memorandum entitled "Clothianidin Registration of Prosper T400 Seed Treatment on Mustard Seed and Poncho/Votivo Seed Treatment on Cotton," November 2, 2010 (see pp. 2, 4). The science that the agency has, and the independent literature find that clothianidin-contaminated pollen and nectar presents an imminent hazard. Because the hazards to honeybee health are present within registered use parameters, it is clear that label changes alone will not offer adequate protection. The issue is not one of application error, in other words. We therefore urge the agency to issue a stop use order immediately. Our nation cannot afford, and the environment cannot tolerate another growing season of clothianindin use.

In addition, because this problem reflects an overuse of the conditional registration program in OPP, we urge you to set an immediate moratorium on the use of such registrations until the program is fully evaluated for compliance with its underlying statutory responsibilities. The conditional registration of clothianidin in 2003 with outstanding data critical to its safety assessment represents a failure that could and should have been avoided. Clearly, the impacts on pollinators were not adequately evaluated prior to the issuance of the conditional registration, despite knowledge of "chronic toxic risk to honey bee larvae and the eventual instability of the hive." This is the case with pollinator protection and a host of other issues that have direct bearing on environmental protection and public health.

In redoing the clothianidin study and evaluating the causes of Colony Collapse Disorder and the larger issue of the pollinator decline crisis, we urge you to establish protocol that fully assesses the complexities that come together to threaten the honeybees. To be fully protective of bees, reviews must consider multiple chemical and cumulative exposures, persistence, and synergistic effects. We can no longer rely on studies of individual chemicals in isolation.

Thank you for your attention to the pollinator crisis and efforts to stem the tide of contamination and poisoning. We look forward to your reply.

Sincerely,

National Honey Bee Advisory Board
Steve Ellis.
Secretary

American Honey Producers Association
Kenneth Haff
President

Pesticide Action Network North America
Heather Pilatic
Co-Director

American Beekeeping Federation
David Mendes
President

Beyond Pesticides
Jay Feldman
Executive Director

Center for Biological Diversity
Justin Augustine
Staff Attorney

55

i want morebooks!

Buy your books fast and straightforward online - at one of world's fastest growing online book stores! Environmentally sound due to Print-on-Demand technologies.

Buy your books online at
www.get-morebooks.com

Kaufen Sie Ihre Bücher schnell und unkompliziert online – auf einer der am schnellsten wachsenden Buchhandelsplattformen weltweit! Dank Print-On-Demand umwelt- und ressourcenschonend produziert.

Bücher schneller online kaufen
www.morebooks.de

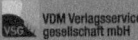 VDM Verlagsservicegesellschaft mbH
Heinrich-Böcking-Str. 6-8   Telefon: +49 681 3720 174   info@vdm-vsg.de
D - 66121 Saarbrücken   Telefax: +49 681 3720 1749   www.vdm-vsg.de

www.ingramcontent.com/pod-product-compliance
Lightning Source LLC
Chambersburg PA
CBHW031547210526
45464CB00003B/1184